Molecular Pathology of Pituitary Adenomas

Molecular Pathology of Pituitary Adenomas

Cristiana Pistol Tanase
Biochemistry/Proteomics Department,
"Victor Babes" National Institute of Pathology,
Bucharest, Romania

Irina Ogrezeanu
Neurosurgery Department,
"D. Bagdasar Arseni" Hospital,
Bucharest, Romania

Corin Badiu
National Institute of Endocrinology,
"C. Davila" University of Medicine and Pharmacy,
Bucharest, Romania

ELSEVIER

AMSTERDAM • BOSTON • HEIDELBERG • LONDON • NEW YORK
OXFORD • PARIS • SAN DIEGO • SAN FRANCISCO • SINGAPORE
SYDNEY • TOKYO

Elsevier
32 Jamestown Road London NW1 7BY
225 Wyman Street, Waltham, MA 02451, USA

First edition 2012

British Library Cataloguing in Publication Data
A catalogue record for this book is available from the British Library

Library of Congress Cataloging-in-Publication Data
A catalog record for this book is available from the Library of Congress

ISBN: 978-0-323-16535-8

For information on all Elsevier publications
visit our website at elsevierdirect.com

This book has been manufactured using Print On Demand technology. Each copy is produced to order and is limited to black ink. The online version of this book will show color figures where appropriate.

Transferred to Digital Printing in 2016

Contents

Preface

Mainly, the book enriches the previous publications regarding pituitary adenomas by promoting information in new fields, such as signal transduction, stem cell markers, microRNAs, "omics" technologies, and less new ones (proliferation, angiogenesis, and apoptosis).

The first point the book focuses on is the classification of pituitary adenomas, taking into consideration clinical functional aspects and aggressiveness. A chapter on pathology emphasizes the relationship between clinical behavior and proliferation markers/rates. The balance between stimulating and inhibiting factors which determine the final angiogenic phenotype of pituitary adenomas, and the molecular systems that regulate the apoptotic process.

A developing field is represented by the key signaling molecules involved in diagnosis, prognosis, and treatment monitoring of pituitary tumors, while a new area targets the stem cells as originators or markers in pituitary adenomas. "Omics" technologies are undertaken as individual or panels of biomarkers. Individual or group signatures of microRNAs can also qualify as potential biomarkers for diagnosis and prognosis of pituitary adenomas. Last, but not least, a short excursion is being made into the current treatment options for pituitary adenomas, that is, their strength and limitations, and the rational behind designing novel therapies, based on releasing hormones, receptors, and other key signaling molecules.

Molecular Pathology of Pituitary Adenomas brings new data about current progress in the understanding of pituitary adenoma pathogenesis and how it impacts upon current attitude. The book provides useful instruments for research and clinical area; a better understanding of tumor biology, the discovery of molecular events at the basis of tumor development was possible due to the modern approach of immunohistochemistry, now routinely used in clinical pathology, and the advanced techniques of electron microscopy, genomics, and proteomics.

The management of pituitary adenomas in the era of evidence-based medicine changed dramatically; new data appear with an increasing speed, and their selection for being implemented in clinical medicine is a continuous challenge.

Molecular Pathology of Pituitary Adenomas represents an update in the field of pituitary tumor research for scientists, clinicians, and everyone dealing with pituitary tumor patients, that is, endocrinologists, internists, oncologists, neurosurgeons, and pathologists.

All authors had equal contributions to this manuscript.

The Authors

Introduction

The molecular pathology of pituitary adenomas has been approached somewhat differently in this book. It describes an update on the levels achieved so far in this field; while classic areas are reviewed rather briefly, new approaches in the pathology of pituitary adenomas are emphasized. Among these, the most important are signaling pathways, stem cells, mRNA, and "omic" technologies as new generic tools for diagnosis, as well as new therapeutic approaches, involving molecular biomarkers. Our own results regarding some of these biomarkers have been inserted in certain chapters.

Special Thanks

Special acknowledgments for the kind invitation of writing this book to editor Dr. Gregory Makowski and publishing director Lisa Tickner. Our warm thoughts go to Professor George Kontogeorgos for encouraging us in our project. We would also like to thank Radu Albulescu for his valuable contribution and critical reviewing, Irina Radu for literature gathering and technical and linguistic assistance, and contributors Lucian Albulescu, Elena Codrici, Linda Maria Cruceru, Monica Livia Gheorghiu, Simona Mihai, Alina Neagoe, Monica Neagu, Adrian Popa, Ionela Daniela Popescu, and Cristina Stancu.

This work was partially supported by projects PNII 62-087/2008 and POS-CCE 212, 685-152/2010.

1 Pituitary Tumor Classification: Functionality, Invasiveness, and Aggressiveness

Introduction

Pituitary tumors, often called *adenomas*, are among the most frequent intracranial tumors (10–15%) after meningiomas and gliomas [1,2]. Their epidemiology shows an incidence in postmortem series between 3.2% and 27%, with an average of 10%. A study performed on more than 3000 autopsied pituitaries showed that the great majority of these tumors are obvious only for pathologists (such tumors are called subclinical), and that fewer than 1/600–1/1000 are macroadenomas [3]. Evaluation using high-resolution imaging, like computed tomography (CT) with contrast for reasons unrelated to pituitary disorders due to the importance of the arterial sequence, shows hypodense lesions in 10–25% [4], while the use of magnetic resonance imaging (MRI) of the brain reveals a pituitary mass in 10% [5] or even less, without contrast (0.3%) [6]. A recent metaanalysis estimated the prevalence of pituitary adenomas at 14.4% in postmortem studies, 22.2% in radiological studies, and 16.9% overall [7].

The new imaging technology changed the way we manage these tumors. A sellar mass can harbor many other lesions besides pituitary adenomas, as presented in Table 1.1.

Since most incidentally discovered pituitary lesions do not have any impact on a subject's health, they are called "incidentalomas." These are lesions which do not affect normal pituitary function and do not compress or invade the surrounding structures. Therefore, they do not require any active medical attitude [5]. From the patient's perspective, knowing about a pituitary adenoma causes anxiety despite normal function and lack of symptoms, while most physicians will consider the pituitary incidentaloma for sequential follow-up, as will be presented below.

Beyond any neoplasia, there is a disturbance of normal cell cycle control in terms of oncogene activation or loss of heterozygosity of tumor suppressor genes. Continuous progress has been made in terms of molecular pathogenesis of pituitary adenomas, as reviewed in the subsequent chapters of this book. Stem cells are a source of continuous tissue renewal, as well as a possible source of monoclonal proliferation under environmental disruptors, aided by a certain genetic background. Recent data suggest that stem cells might be involved in this process [8,9].

From their epidemiology to their molecular aspects, pituitary adenomas represent a broad spectrum of disorders that can be analyzed and classified according to autonomous secretion, clinical aspects in diagnosis and treatment, pathology in

Molecular Pathology of Pituitary Adenomas. DOI: 10.1016/B978-0-12-415830-6.00001-9

Table 1.1 Pituitary Masses

Physiologic Hyperplasia	Pituitary Tumors	Malignancies	Cysts	Others
Lactotroph in pregnancy Thyrotroph in myxedema Somatotroph hyperplasia (ectopic GH-releasing hormone, GHRH) Gonadotroph (arguable) Corticotroph (arguable)	Pituitary adenomas (most frequent) Craniopharingioma Meningioma	Ectopic pinealoma (germinoma) Granulomatous diseases Sarcoma, chordoma Pituitary carcinoma—rare Metastases (lung, breast)	Rathke's cleft Dermoid Arachnoid	Pituitary abscess Pituitary tuberculosis Lymphocytic hypophysitis Vascular abnormalities

Table 1.2 Clinical Functional Classification

Pituitary Adenoma	Clinical Signs	Gross Prevalence	Comments
Prolactinoma	Amenorrhea–galactorrhea in women	35%	In men, occur as nonfunctioning adenomas
Acromegaly	Physical changes	20%	Onset much before diagnosis
Cushing's disease	Clinical Cushing's	13%	In adults, more frequent than in children
Thyrotropinoma	Pituitary thyrotoxicosis	2%	TSH inadequate high for increased T3, T4
Nonfunctioning adenoma	Hypopituitarism	30%	Can include prolactinoma in men, gonadotroph adenomas, or other silent adenomas

terms of light microscopy (LM) and electron microscopy (EM) features, and special issues, such as the clinical impact of aggressive pituitary adenomas or rare pituitary carcinoma cases. Despite being histologically benign, they can be severe and life-threatening due to local invasion and compression, metabolic, or cardiovascular complications (Table 1.2).

The World Health Organization (WHO) publication "Histological Typing of Endocrine Tumours" uses a five-tier classification in which endocrine activity, imaging, operative findings, and detailed pathology are integrated [10]. According to the mentioned classification, there are only three accepted types of anterior pituitary lesions: typical pituitary adenoma, atypical pituitary adenoma, and pituitary carcinoma. A step forward in classifying pituitary tumors was the use of tumor markers after 2005–2006. The Ki-67 labeling index (LI) is widely used due to its correlation with invasiveness, and probably prognosis as well. While adenomas showing increased (>3%) LI and extensive p53 immunoreactivity are considered "atypical adenomas," suggesting an aggressive potential or malignant transformation, the term *pituitary carcinoma* is applied exclusively when cerebrospinal and/or systemic metastases are identified [11].

As most of the tumors are subclinical and they never get removed by the neurosurgeon, it is important to focus on required medical attitude and dividing them into incidentalomas, which do not require treatment, and pituitary adenomas, which impose detailed diagnosis and active pharmacological, surgical, or radiation therapy. The variation in pathology, despite significant progress in this area, seldom influences the clinical decision in a significant manner [12].

The wide variation in human resources and technical facilities, as well as financial constraints, can affect the high-detail characterization (genetic and molecular level) of pituitary tumors. However, the clinical and basic research pituitary community can bring together resources in difficult, rare cases, ensuring an adequate clinical approach as well as research material.

Clinical Presentation and Classification

Upon clinical presentation, pituitary adenomas can be classified as secreting tumors (such as growth hormone (GH), adrenocorticotropic hormone (ACTH), prolactin (PRL) in women, or the rare thyroid-stimulating hormone (TSH)–secreting adenomas) or clinically nonfunctional adenomas. Tumor secretion is autonomous, which triggers various clinical syndromes and allows for appropriate testing. In addition to tumor secretion, pituitary adenomas might compress the neighboring structures, with optic chiasma or cavernous sinus syndrome, sphenoid sinus invasion or extension toward the base of the brain, all included in the mass effect. Last but not least, normal pituitary function can be impaired, leading to various degrees of pituitary failure.

The most common adenomas (30–35%) are PRL-secreting tumors. GH-secreting adenomas cause acromegaly and gigantism. Less common are ACTH-secreting adenomas, which cause Cushing's disease and TSH-secreting tumors, triggering pituitary hyperthyroidism. The remaining pituitary adenomas, representing approximately a third, are clinically silent and are known as nonfunctioning pituitary adenomas (NFPAs), meaning that they only cause symptoms due to tumor growth. However, the last category is subdivided after detailed immunopathology into gonadotropinomas, tumors with secretion of "mute" hormones, and null cell adenomas, which are devoid of immunoreactivity for classic anterior pituitary hormones.

Prolactinomas

Prolactinomas are adenomas associated with increased PRL levels usually above 100 ng/ml; the serum PRL levels usually correlate with the tumor size [13]. PRL-producing tumors show the highest incidence among pituitary adenomas in childhood and adolescence [14]. They represent the most common neoplasm of the anterior pituitary in clinical series and in autopsy material [3]. Although they were among the first type of pituitary adenomas included in surgical series three decades ago, their operating frequency has decreased dramatically after the introduction of dopamine agonist therapy, which has proved to be more effective.

Owing to clinical manifestation mostly presenting as amenorrhea–galactorrhea, PRL-secreting adenomas are usually microadenomas in females. Due to the high frequency of hyperprolactinemia as isolated endocrine disorder on one hand and pituitary microincidentaloma on the other, it is sometimes difficult to sustain the diagnosis of microprolactinoma. PRL might be increased due to drugs (such as neuroleptics, antidepressants, estrogens, and metoclopramids), pituitary stalk syndrome, local thoracic or breast lesions, or even postpartum or postabortum hyperprolactinemia [15]. High PRL levels without any effect upon gonadal function can be observed in macroprolactinemia [16]. In contrast, they present in males as macroadenomas in the majority of cases because of delayed diagnosis due to their modest clinical symptoms, such as decreased libido and sexual dysfunction [13].

Clinical Syndrome of Prolactinomas

The clinical syndrome of prolactinomas is highly suggestive when amenorrhea and galactorrhea occur in women, as Chiari in 1832 and Fromel in 1882 had described [17]. Disorders of ovulation involve PRL more than in the frame of prolactinomas, higher PRL levels being attributed to many disorders [18] (Table 1.3). Pituitary adenomas can be associated with hyperprolactinemia (HPRL), either by tumor-autonomous secretion of PRL (prolactinomas, acromegaly) or by lack of inhibitory control in the macroadenomas with suprasellar extension (SSE) and pituitary stalk syndrome [19]. In addition, HPRL can be triggered by drugs that interfere with dopaminergic control, decreased removal in kidney failure, as well as reflex mechanisms [20].

Sometimes galactorrhea is obvious, spontaneous, and in high amounts, while it might be present only upon careful breast clinical examination, and sometimes it may even be absent. Although most prolactinomas occur in adults, puberty can be delayed and associated with primary amenorrhea in a childhood-onset prolactinoma. While prolactinomas are microadenomas or macroadenomas in women, they are mostly macroadenomas in men, therefore being associated with low libido and impotence, mass lesions (optic chiasma, cavernous sinus syndrome), and global hypopituitarism. Rarely, galactorrhea may occur in men, but this is related more to the degree of breast development due to estrogens than to PRL.

Hyperprolactinemia is defined as a fasting PRL level above 20 ng/ml in men and 25 ng/ml in women, using immunoassays. Various methods have been used in PRL assays. Compared with the first competitive radioimmunoassay, the introduction of the two-site monoclonal "sandwich" assays, immunoradiometric assay (IRMA) and

Table 1.3 Causes of Hyperprolactinemia

Physiologic	Pituitary Disorders	CNS Disorders	Systemic Diseases	Drugs
Pregnancy	Prolactinomas	Tumors	Severe hypothyroidism	Antidepressants Valproic acid
Breastfeeding	Mixed pituitary adenomas	Granulomatous diseases	Liver cirrhosis	Anxiolytics
Breast stimulation	Acromegaly	Vascular disorders	Chronic renal failure	Selective serotonin reuptake inhibitors
Sleep	Nonfunctioning adenomas	Autoimmune disorders	Polycystic ovary syndrome	Antihypertensives Methyldopa, Reserpine, Verapamil
Stress	Pituitary stalk syndrome	Hypothalamic tumors or metastasis	Estrogen-secreting tumors	Antipsychotics H2 receptor blockers Metoclopramide
	Lymphoid hypophysitis Cushing's disease	Cranial irradiation	Pseudocyesis	Estroprogestins
	Empty sella syndrome	Seizures	Chest wall trauma Herpes zooster	Phenotiazines and Opiates

immunochemiluminescent assays (ICMA), has been associated with more rapid and accurate hormone measurement. In these assays, two different antibodies, one attached to a solid surface (capture antibody) and the other added with the patient's serum sample (signal antibody), label the hormone (antigen) with a tracer substance. After the unbound signal antibodies are washed away, the remaining tracer detected correlates to the "level" of PRL. However, when there is an excess of antigen, the soluble antigen binds separately to each antibody, giving rise to false low values (the "hook" effect) despite significant clinical syndrome [21]. On the other hand, if the patient has high levels of big-big PRL or macroprolactinemia, high PRL levels are encountered that are devoid of any clinical impact on the reproductive axis [16].

GH-Producing Adenomas

GH-producing adenomas represent 20% of pituitary tumors in surgical material, and they show no sex predilection. Their lower incidence in clinical series is due to the inclusion of medically treated adenomas. The prevalence is between 40 and 80 cases per million, with an annual incidence of 3–4 new cases per million, but newer studies suggest an increased prevalence [22]. Most tumors are macroadenomas producing ballooning enlargement of the *sella turcica*, often with upward and/or lateral extension. They are characterized by GH/IGF1 (insulin-like growth factor 1) excess and are associated clinically with either gigantism or acromegaly, depending on the patient's age at the onset of the disease. Additional clinical associations include mass effects and tumor-induced anterior pituitary failure.

Clinical Syndrome in Acromegaly

Acromegaly defines a clinical syndrome of "Acros Megalos," meaning increased extremities. The disease was described in Egyptian (Akhenaten) and Roman (Maximilian Thrax) antiquity, but under this name, it was proposed by the French neurologist Pierre Marie in 1885. Due to progressive somatic changes, the onset of the disease is insidious, and over many years, the patient observes coarse facial features associated with dental malocclusion, increased size of hands and feet, high blood pressure, heart failure, sleep apnea, and a diminished temporal visual field. In addition, amenorrhea might occur in women and decreased libido and/or impotence in men. Occasionally, galactorrhea might occur due to PRL cosecretion. Joints are also affected, leading to degenerative changes, mostly in the coxofemural and knees, which cause severe motility dysfunction. The liver, heart, and thyroid increase in size (visceromegaly), reflecting the same excess of GH. Increased body weight and body mass index (BMI) is not relevant for defining obesity since often the adipose tissue mass is not increased [23]. Mammosomatotroph adenomas represent the most common cause of gigantism, when they occur in young patients, before the epiphyses are closed.

The global health impact of acromegaly results in increased cardiovascular mortality at two to four times the rate in the general population.

ACTH-Producing Adenomas

Corticotroph adenomas are the most frequent form (two-thirds) of Cushing syndrome in adults, leading to the clinical and biochemical features known as Cushing's disease. The features were described by Harvey Cushing in 1932 [24]. Prolonged overproduction of cortisol and other adrenocorticoids include centripetal obesity, bruising, plethoric facies, red striae, muscle weakness, and backache due to osteoporosis. These are related to protein catabolism due to excess glucocorticoids. In addition, acne and hirsutism are the expression of excess androgens, while metabolic actions are steroid diabetes, dyslipidemia, and hypertension. Excess steroids can block other pituitary hormones such as gonadotrophins, leading to infertility and menstrual disorders, nonspecific TSH suppression, or GH leading to dwarfism in children. Seldom in Cushing's disease, but often in ectopic ACTH secretion, hyperpigmentation due to ACTH overproduction can occur.

However, the clinical features can be triggered by a pituitary corticotroph adenoma, an adrenal tumor, or an ectopic, ACTH-producing tumor. The laboratory workup should end up with the clear etiology, showing the surgeon the lesion to be removed in order to reestablish a normal adrenal axis. Dexamethasone suppression tests and high-resolution pituitary and adrenal imaging are compulsory.

The ACTH-producing adenoma derives from the anterior pituitary corticotrophs that synthesize several additional proopiomelanocortin-cleaved peptides, including β-lipotrophin and β-endorphin [25]. Corticotroph adenomas represent approximately 10% of pituitary tumors. Corticotroph adenomas show female predilection, with an 8:1 female-to-male ratio [23], and they are rare in childhood. The substantial majority of adenomas associated with Cushing's disease measure less than 1 mm; pituitary MRIs reveal a microadenoma in about 80% of cases. However, microadenomas less than 4 mm are not demonstrated by imaging procedures, even though there is clinical evidence of adenoma. In addition, histology cannot identify a small subset of tumors due to loss during surgery or during sectioning of the tissue specimen.

TSH-Producing Adenomas

TSH-producing adenomas are rare, representing approximately 1% of pituitary adenomas. Most tumors are macroadenomas and show female predilection [26]. Thyrotroph adenomas mostly secrete TSH in excess and present with a diffuse goiter and hyperthyroidism. The biochemical picture is characterized by inappropriate TSH secretion, while high levels of thyroid hormones are found. The TRH challenge test (400 μg i.v.) is not able to stimulate and T3 administered at 75 μg/day for 5 days is not able to suppress the TSH levels, showing the autonomous character of tumor secretion. However, somatostatin analogs are able to suppress the TSH secretion and are being used as assays as well as therapy. Misdiagnosed patients treated with surgery or radioiodine often exhibit progressive growth and invasiveness of the thyrotroph adenoma.

Clinically Nonfunctioning Adenomas

NFPAs are tumors without a clinical evidence of hormone production. However, they attract medical attention due to a high-resolution CT/MRI scan and mass effects, as optic chiasma or cavernous sinus syndrome associated with various degrees of pituitary failure. Microadenomas (<1 cm average diameter) can be incidental findings of CT or MRI scans [5]. If the normal pattern of pituitary secretion is not changed at basal and dynamic testing, and if at serial evaluation (at 6 months, 1 year, 3 years, and 5 years), the tumor does not increase in size, there is no need for treatment and the tumor is called incidentaloma [5]. Therefore, incidental imaging and finding of a pituitary mass is not similar to pituitary incidentaloma diagnosis since a macroadenoma can be an incidental finding and sometimes pituitary adenomas in men can go undiagnosed for many years. Rarely, such a tumor can exhibit a rapid increase in size, with signs of pituitary failure and mass effects, showing a pituitary apoplexy and requiring a rapid neurosurgical approach.

Clinically nonfunctioning adenomas can hide a "mute" hormone secretion, as is the case of prolactinomas in men, gonadotroph adenomas, silent ACTH, TSH, or even GH. These subtypes are revealed only if immunohistochemical staining is performed. If the tumor is negative in response to routine immunostaining, then it is called "null cell" adenoma.

Aggressive Pituitary Tumors

In general, all aggressive tumors originating anywhere in the body can invade adjacent structures. They may also invade blood vessels and lymphatics, resulting in the development of distant metastases. Certain aggressive tumors not only may respond poorly to treatment but also show a distinct tendency to recur even after initially successful treatment. When a tumor becomes aggressive, various histopathology and molecular marker changes become apparent. These signs of aggression include increasing cellular atypia, vascular proliferation and markers of angiogenesis, hemorrhage, necrosis, increased signs of proliferation, and an overexpression of various components of transcription.

When pituitary tumors become aggressive, their clinical behavior mimics that of other aggressive tumors. They may invade surrounding structures such as the cavernous sinus, sphenoid sinus, other nearby bony structures, and the central nervous system (CNS). The tumors may be resistant to medical treatments such as dopamine or somatostatin agonists. After initial surgical resection, they may recur locally and subsequently invade the tissues surrounding the sellar region. In rare instances, they may even metastasize to distant sites (cerebral or extracerebral), at which point they are deemed pituitary "carcinomas."

However, while they show the same clinical behavior, they do not seem to necessarily follow the histological and molecular patterns that are observed in other aggressive neoplasms. Even those that are metastasized may not show the typical histological signs of tumor aggression. Researchers have investigated measures of cell proliferation, cell cycling, cell-to-cell adhesion, alterations of extracellular matrix components, growth factor signaling, and angiogenesis in pituitary tumors. Investigations have documented that many of the markers of these pathways are overexpressed in aggressive tumors elsewhere in the body, and the role of these markers in pituitary tumors has also been studied. The results are mixed, in that the expression of some markers correlates with the aggressiveness of pituitary tumors while that of others does not.

Pituitary Carcinomas

Morphologic separation of pituitary adenomas from a carcinoma is not possible. The term *pituitary carcinoma* is applied exclusively when cerebrospinal and/or systemic metastases are definite. Primary pituitary carcinomas are very rare, representing approximately 0.2% in surgical series. Pituitary carcinomas affect adults only, with no age or sex preponderance.

They are associated with poor prognosis; the mean survival rate is 8 years for approximately 80% of patients with documented metastases. Two-thirds of pituitary carcinomas are endocrinologically functional, primarily producing PRL or ACTH. Metastases of pituitary carcinomas include dissemination throughout the subarachnoid space and lymphatics. Hematogenous dissemination has been reported only in a few cases, primarily of ACTH-producing carcinomas.

Even though they are monoclonal in nature, pituitary tumors prove functional diversity and behavioral unpredictability. Therefore, much effort was done in order to evaluate tumorigenesis-inducing factors and their correlation with clinical–pathological studies. In fact, pituitary tumorigenesis is a complex interplay between genetic and humoral factors, in which alterations occur in different signaling pathways that link the extracellular milieu with the nucleus and govern tumor formation.

Pituitary Incidentaloma

The diagnosis of pituitary incidentaloma relates to the anatomic evidence of a pituitary adenoma, while no complaints or clinical or laboratory abnormalities can be found when a standard pituitary evaluation is performed [5]. Most of them are microadenomas, at an average of 10%, since in 18,902 cases in autopsy series, only seven were above 1 cm (0.03%). They are equally distributed in men and women and throughout all age groups from 18 to 70 years. Contrast high-resolution CT showed an incidence of 10%, while cranial 1.5 T MRIs [6] showed only 0.3%.

The clinical attitude in the face of a pituitary tumor should be based upon several items:

1. Obvious clinical syndrome of prolactinoma, acromegaly, Cushing, or showing tumor-autonomous secretion (PRL, GH, or ACTH); this evaluation of pituitary hyperfunction should include basal PRL; IGF1, GH during Oral glucose tolerance test (OGTT) OGTT; basal, midnight, and after overnight, 1 mg dexamethasone suppression plasma cortisol levels.
2. Pituitary failure on the gonadal, thyroid, or adrenal axis; for women until their fifties, progesterone levels at the twenty-first day and E2, follicle-stimulating hormone (FSH), luteinizing hormone (LH) early during menstrual cycle are good indicators of ovarian/ovulatory function, while a failure of FSH and a rise in LH after menopause denote pituitary failure without any need of a dynamic test. In men, a testosterone and spermogram indicate gonadal axis status. Evaluation of thyroid function should be based upon fT4, while TSH is not relevant in central hypopituitarism. The adrenal axis can be evaluated either by a short ACTH stimulation test or by insulin-induced hypoglycemia. Evaluation for hypopituitarism is compulsory in cases of macroadenomas, while this is optional in microadenomas.
3. Compressive signs on optic pathway or cavernous sinus invasion; despite lack of symptoms, an objective evaluation (Goldman perimetry) showing the normal or narrowed visual field, as well an optic fundus showing a normal or pale papilla, are required.
4. Complications with metabolic impact (secondary diabetes mellitus, dyslipidemia), cardiovascular impact (high blood pressure, heart hypertrophy, or dilatative cardiomyopathy).
5. Laboratory findings supporting the clinical suspicion, in terms of dynamic tests showing lack of suppression and partial or global pituitary failure.

If all of the above are negative, then a clinically nonfunctioning adenoma is documented. In order to state the diagnosis of incidentaloma, it is necessary to prove the absence of tumor evolution or clinical/biochemical pituitary dysfunction over time. The recommended timing of evaluation is at 6 months, 1 year, 2 years, and 5 years (Figure 1.1). The decisive distinction between incidentaloma and pituitary adenoma is that no treatment is needed for incidentaloma.

Radiological Classification of Pituitary Adenomas

The evaluation of pituitary adenomas includes a sellar region imaging procedure, in addition to a clinical and biochemical approach. Among the first used was the

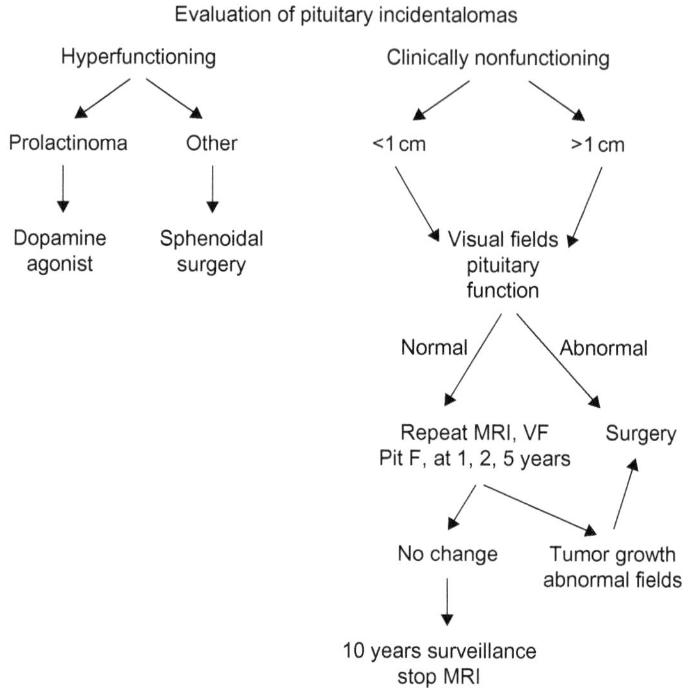

Figure 1.1 Pituitary adenoma evaluation algorithm. The clinical and biochemical syndrome of autonomous secretion allows an appropriate attitude. In case of a documented nonfunctioning adenoma, with no impact upon neighboring structures or pituitary function, a "wait-and-see policy" is preferable, with sequential MRI follow-up.

anatomo-radiologic classification of the *sella turcica* alterations in four grades stated by Hardy [27], considering the sellar X-ray. Grade 0 was without any sellar erosion; grade 1 with local erosion due to a microadenoma (tumor size <1 cm diameter); grade 2 macroadenomas, with or without extrasellar expansion; grade 3 macroadenomas with invasion of sellar bone; the last grade, 4, was big tumors, always macroadenomas with invasion and erosion of the sphenoid bone. The extension of these tumors could be SSE, toward the optic pathway, parasellar, toward the cavernous sinus, or inferior, toward the sphenoid sinus. Giant adenomas, with an average diameter above 4 cm, could expand in all directions.

Another useful classification is related to the lateral extension toward the cavernous sinus and carotid arteries, known as the Knosp classification (Figure 1.2). Grade 0 represents the normal condition of the cavernous sinus space. The adenoma does not pass the tangent of medical aspects of supra- and intercavernous internal carotid artery (ICA). In grade I, the medial tangent is passed, but the extension does not go beyond the intercarotid line, which is the line drawn between the cross-sectional centers of the intra- and supracavernous ICA. Grade II is characterized by the tumor extending beyond the intercarotid line but not extending beyond or tangent to the lateral aspects

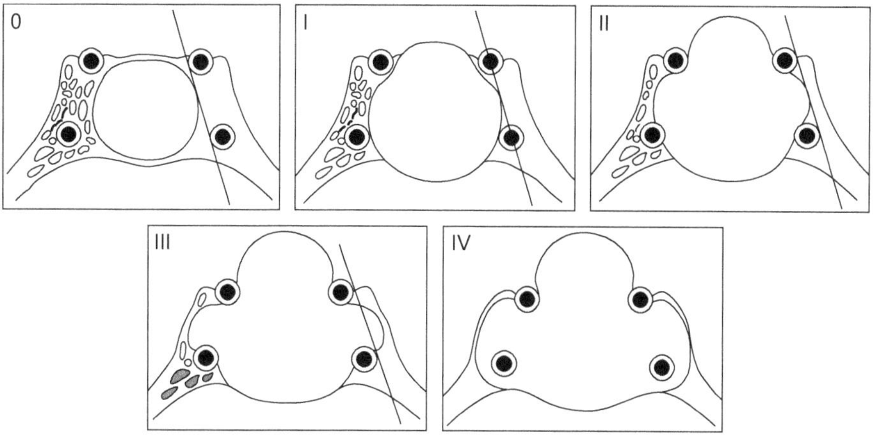

Figure 1.2 Knosp classification. Grade 0 represents the normal anatomy of the cavernous sinus space. The adenoma does not extend over the tangent of medial aspects of supra- and intercavernous ICA. In grade I, the medial tangent is crossed, but the extension does not go over the intercarotid line, drawn between the cross-sectional centers of the intra- and supracavernous ICA. Grade II is characterized by the tumor extending outside the intercarotid line, but not beyond or tangent to the lateral aspects of the intra- and supracavernous ICA. Grade III is a tumor extending laterally to the lateral tangent of the intra- and supracavernous ICA. Grade IV is characterized by total inclusion of the ICA by the adenoma.

of the intra- and supracavernous ICA. Grade III is characterized by the tumor extending laterally to the lateral tangent of the intra- and supracavernous ICA. Grade IV is characterized by total encasement of the intracavernous carotid artery [28].

In the modern years of advanced imaging technology, sellar X-ray is of history or seldom, screening. The high-resolution CT, with contrast enhancement and MRI, is based upon microadenomas (<1 cm), mezoadenomas (≈1 cm), macroadenomas (>1 cm), and giant adenomas (>4 cm). However, only tumor size does not correlate with Ki-67 LI or with vascular invasion markers like vascular endothelial growth factor (VEGF), showing a long evolution time [29].

Pituitary Apoplexy

Definition

Described in 1898 by Bailey [30], pituitary apoplexy is a rare but life-threatening medical emergency. It is a clinical syndrome characterized by the sudden onset of headache, vomiting, visual impairment, decreased consciousness, and impaired pituitary function caused by hemorrhage and/or infarction of the pituitary gland.

The symptoms are caused by the rapid enlargement of a pituitary adenoma, usually due to the hemorrhagic infarction of the tumor, and evolve within hours or days. Pituitary apoplexy occurs most often in a previously undiagnosed clinically nonfunctioning macroadenoma, but it has been described in patients with secreting adenomas as well.

The term *pituitary apoplexy* should be used only for the clinically overt syndrome. There has been also described asymptomatic pituitary hemorrhage and/or infarction ("subclinical pituitary apoplexy"), which may be detected on routine imaging or during histopathological examination in 25–28% of patients with pituitary adenomas [31].

Incidence

The incidence of apoplexy in pituitary adenomas is between 2% and 7% when defined on the basis of clinical signs and surgical or histopathological evidence. In more than 80% of patients, pituitary apoplexy is the first presentation of the underlying pituitary tumor. Most patients present in their fifties or sixties, and there is a slight male preponderance [32].

Since the clinical picture may mimic subarachnoid hemorrhage, bacterial meningitis, or stroke, patients may present in different specialties and the diagnosis can be difficult and delayed.

Precipitating factors have been identified in up to 40% of cases of pituitary apoplexy (Table 1.4), the most common factor being hypertension [33].

Clinical Manifestations

Hemorrhage induces a rapid increase of the pituitary adenoma, with an increase in the intrasellar pressure and compression on the SSE and lateral structures (optic chiasm, cavernous sinuses). The initial symptom in up to 100% of patients is sudden, severe *headache*, which may be accompanied by *nausea and vomiting. Visual disturbances* may occur in up to 75% of patients and include visual loss, ophthalmoplegia, paresis of third, fourth, and/or sixth cranial nerves, decreased visual acuity, and visual field defects, specifically bitemporal hemianopia. Extravasation of blood or necrotic tissue into the subarachnoid space can cause *meningism* with fever, photophobia, and altered consciousness levels. Rarely, *cerebral ischemia* can result from

Table 1.4 Precipitating Factors for Pituitary Apoplexy

Hypertension
Cardiac surgery (e.g., coronary artery bypass grafting)
Anticoagulation (initiation or withdrawal)
Closed head trauma
History of pituitary irradiation
Coagulopathies
Treatment with dopamine agonists, estrogens, or gonadotrophin-releasing hormone (GnRH) agonists
Pituitary stimulation testing (TRH, Corticotrophin releasing hormone (CRH), GnRH, insulin)
Major surgery
Diabetes mellitus
Pregnancy

either mechanical compression of the carotid artery against the anterior clinoid or vasospasm secondary to subarachnoid hemorrhage.

Partial or complete *hypopituitarism* is present in up to 80% of patients. Usually, there is acute *secondary adrenocortical insufficiency* (in about 70% of patients), with hypotension, nausea, and vomiting. Although *thyrotrophin and gonadotrophin deficiencies* are observed in 50% and 75% of patients, respectively, they may not be clinically expressed unless they have preceded the apoplectic episode. *Diabetes insipidus* may occur transiently in about 4% of patients with apoplexy and persistently in 2% of these patients [34–36].

Evaluation

All patients with suspected pituitary apoplexy should have urgent blood samples drawn to check electrolytes, renal function, liver function, clotting screen, full blood count, and random cortisol (before i.v. hydrocortisone is initiated), fT4, TSH, IGF1, GH, LH, FSH, PRL, and, in addition, testosterone in men and estradiol in women. Hyponatremia has been reported in up to 40% of patients because of the syndrome of inappropriate antidiuretic hormone secretion and/or hypocortisolism [32].

Urgent MRI should be done because it confirms the diagnosis of pituitary apoplexy in more than 90% of the patients as compared to CT, which identifies sellar masses in over 80% of patients but diagnoses pituitary apoplexy only in about 25% of them [34,37]. T1-weighted MR images usually show areas of high signal intensity, while T2-weighted MR images show areas of low signal intensity, either large, spots, or rim (Figure 1.3). MRI and MR angiogram techniques may help to differentiate an aneurysm from pituitary apoplexy.

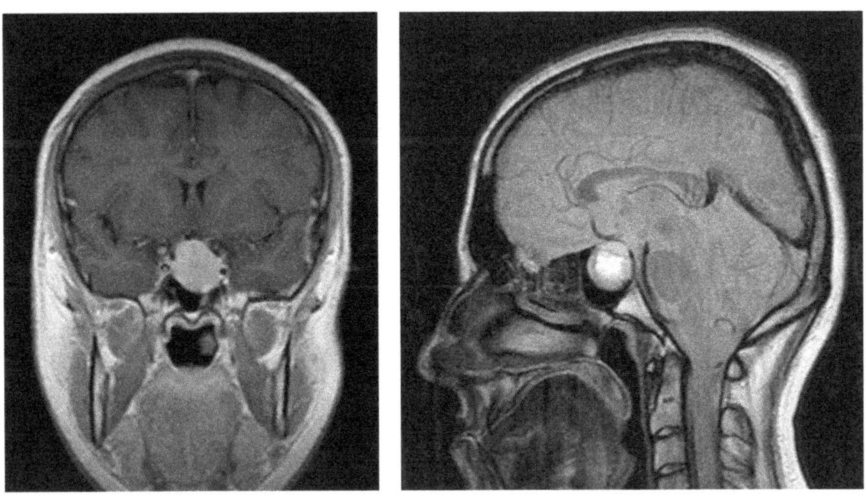

Figure 1.3 Pituitary tumor apoplexy. *Left*: Coronal T1-weighted MRI, showing a pituitary macroadenoma with SSE and invasion of the left cavernous sinus. *Right*: Sagittal T1-weighted MRI with high signal in the pituitary adenoma mass, suggesting pituitary apoplexy.

A formal ophthalmologic assessment for visual acuity and for visual fields, using Humphrey visual field analyzer or Goldmann perimeter, must be undertaken when the patient is clinically stable, preferably during the first 24 h.

Management

Patients should be referred to a multidisciplinary team comprising an endocrinologist, a neurosurgeon, and an ophthalmologist, among others. Other than patients with worsening neurological symptoms in whom surgery is indicated, it is currently unclear whether conservative or surgical management carries the best outcome.

Initial medical management should include careful *assessment of fluid and electrolyte balance*, *replacement of corticosteroids*, and supportive measures to *ensure hemodynamic stability*, as recommended by the recently published UK guidelines for the management of pituitary apoplexy [32]. In patients with hemodynamic instability, altered consciousness level, reduced visual acuity, and severe visual field, as well as in patients with overt adrenal insufficiency or low morning serum cortisol, corticosteroid therapy should be initiated, with hydrocortisone 100–200 mg as an i.v. bolus, followed either by 20–40 mg/h by continuous i.v. infusion (preferably) or by 50–100 mg for every 6 h by i.m. injection.

After the acute episode, the hydrocortisone dose should be tapered to a maintenance dose of 20–30 mg/day, orally usually in three divided doses. Dexamethasone may be used in the acute phase to reduce brain edema, but it is not recommended as a long-term steroid replacement therapy. Reassessment of the pituitary ACTH reserve should be done 1–3 months after the acute pituitary apoplexy.

Surgery should be considered for patients with severe neuro-ophthalmic signs (e.g., severely reduced visual acuity and severe and persistent or deteriorating visual field defects) or deteriorating level of consciousness. A semi-elective transsphenoidal surgery is preferable, performed by an experienced pituitary surgeon (with at least five or more transsphenoidal pituitary operations per year), rather than by on-call neurosurgical team. Surgery should be performed preferably within the first week of the onset of symptoms. Early decompression has been shown in a few observational studies to improve the endocrine and visual outcome. However, other recent studies have suggested that the outcome was similar in patients managed conservatively or by early surgical intervention [34,36,38,39]. These studies have been criticized for selection bias, patients treated conservatively having milder neuro-ophthalmic signs [32].

Conservative management may be indicated in patients with pituitary apoplexy who have no or mild and stable neuro-ophthalmic signs. There are no evidence-based criteria to justify the clinical decision between a conservative approach and neurosurgical intervention in these patients. Mild neuro-ophthalmic signs tend to improve spontaneously in most patients [36–39]. Ocular paresis (due to involvement of III, IV, or VI cranial nerves), in the absence of visual field defects or reduced visual acuity, is not in itself an indication of immediate surgery, as resolution usually occurs within days or weeks with conservative management. It has been suggested that a large hypodense area within the tumor (as opposed to several small hypodense areas) was associated with subsequent tumor resolution [40].

However, due to the unpredictable evolution of the pituitary apoplexy, patients treated conservatively should be closely monitored (frequent clinical, neurological, and ophthalmologic evaluation), and surgical intervention must be considered if neuro-ophthalmic or general signs fail to improve or deteriorate.

Postoperative assessment should include evaluation of the pituitary function after 2–3 days and again after 4–8 weeks, as well as ophthalmologic assessment. Transient diabetes insipidus is apparent postoperatively in up to 16% of the patients with pituitary apoplexy during their hospital stay and may be permanent in some of them [34]. Other potential postoperative complications include pituitary insufficiency deficiency, visual loss, cerebrospinal fluid leakage, and meningitis.

Long-Term Outcome

Partial or complete recovery of pituitary function occurs in up to 50% of patients. Patients with low serum PRL levels due to very high intrasellar pressure rarely recover from hypopituitarism after decompressive surgery [41].

Early studies showed a better endocrine outcome in surgically managed patients, while more recent retrospective studies found no statistically significant differences between the surgically and conservatively managed patients [38,39]. Nearly 80% of the patients will need some form of hormone replacement after apoplexy. GH deficiency is the most frequent deficit after apoplexy (in almost all patients), followed by ACTH (in 60–80% of patients), TSH (in 50–60%), vasopressin (in 10–25%), and testosterone (in 60–80% of men) [32].

Visual acuity, visual field defects, and ophthalmoplegia have been reported to improve in the majority of the patients (70–95%) after surgical decompression (especially in those operated within 7 days of the appearance of these symptoms) [34,35], but also in conservatively managed patients [37,39]. The improvement occurs in the immediate postoperative period and may continue for several weeks after surgery. Visual recovery is lower in patients presenting with monocular or binocular blindness (improvements in about 70% and 45%, respectively) [42].

Cure of hormone secreting or nonsecreting pituitary tumors has also been reported after pituitary apoplexy. Clinical or subclinical apoplexy has induced spontaneous remission of acromegaly in more than 20 patients [43,44]. It may lead to intrasellar calcifications [45]. Pituitary apoplexy was associated with long-term spontaneous remission in a few patients with Cushing's disease [46]; although prolactinomas frequently show tumor hemorrhage, necrosis, or calcifications, cure after pituitary apoplexy seems to be unusual in these tumors. Apoplexy in microadenomas is an extremely rare event [47]. Currently, there is no indication for surgery in pituitary incidentalomas in order to prevent apoplexy.

Monitoring

Postapoplexy, all patients should be carefully monitored for recurrence of tumor growth, which has been documented in surgical, as well as in conservatively managed, patients. However, recent studies have not shown significant differences between these groups [36,39]. Of note, recurrence has been described up to 18 years following the acute episode [43]. An MRI scan is recommended at 3–6 months after apoplexy, and thereafter an annual MRI scan should be considered for the next 5 years, then every 2 years [32]. Since hypopituitarism is persistent in most of the patients, they require lifelong endocrine follow-up.

References

1. Saeger, W., Ludecke, D.K., Buchfelder, M., Fahlbusch, R., Quabbe, H.J. and Petersenn, S. (2007) Pathohistological classification of pituitary tumors: 10 years of experience with the German Pituitary Tumor Registry. *Eur. J. Endocrinol.*, 156: 203–16.
2. Melmed, S. (2008) Update in pituitary disease. *J. Clin. Endocrinol. Metab.*, 93: 331–38.
3. Buurman, H. and Saeger, W. (2006) Subclinical adenomas in postmortem pituitaries: Classification and correlations to clinical data. *Eur. J. Endocrinol.*, 154: 753–58.
4. Chidiac, R.M. and Aron, D.C. (1997) Incidentalomas. A disease of modern technology. *Endocrinol. Metab. Clin. North Am.*, 26: 233–53.
5. Molitch, M.E. (2009) Pituitary tumours: Pituitary incidentalomas. *Best. Pract. Res. Clin. Endocrinol. Metab.*, 23: 667–75.
6. Vernooij, M.W., Ikram, M.A., Tanghe, H.L., Vincent, A.J., Hofman, A., Krestin, G.P., et al. (2007) Incidental findings on brain MRI in the general population. *N. Engl. J. Med.*, 357: 1821–28.
7. Ezzat, S., Asa, S.L., Couldwell, W.T., Barr, C.E., Dodge, W.E., Vance, M.L., et al. (2004) The prevalence of pituitary adenomas: A systematic review. *Cancer*, 101: 613–19.
8. Levy, A. (2008). Stem cells, hormones, and pituitary adenomas. *J. Neuroendocrinol.*, 20, 139–40.
9. Xu, Q., Yuan, X., Tunici, P., Liu, G., Fan, X., Xu, M., et al. (2009) Isolation of tumour stem-like cells from benign tumours. *Br. J. Cancer*, 101: 303–11.
10. Scheithauer, B., Horvath, E., Kovacs, K. and Lloyd, R.V. (2004) Tumours of the adenohypophysis. In *International Histological Classification of Tumours: Histological Typing of Endocrine Tumours* (E. Solcia, G. Klöppel and L.H. Sobin, eds.), pp. 15–28. Berlin: Springer-Verlag.
11. Asa, S.L. and Ezzat, S. (2009) The pathogenesis of pituitary tumors. *Annu. Rev. Pathol.*, 4: 97–126.
12. Perry, A. and Scheithauer, B.W. (2006) Commentary: Classification and grading of pituitary tumors. Observations of two working neuropathologists. *Acta Neuropathol.*, 111: 68–70.
13. Colao, A., Vitale, G., Cappabianca, P., Briganti, F., Ciccarelli, A., De, R.M., et al. (2004) Outcome of cabergoline treatment in men with prolactinoma: Effects of a 24-month treatment on prolactin levels, tumor mass, recovery of pituitary function, and semen analysis. *J. Clin. Endocrinol. Metab.*, 89: 1704–11.
14. Colao, A., Loche, S., Cappa, M., Di, S.A., Landi, M.L., Sarnacchiaro, F., et al. (1998) Prolactinomas in children and adolescents. Clinical presentation and long-term follow-up. *J. Clin. Endocrinol. Metab.*, 83: 2777–80.
15. D'Ercole, M., Della Pepa, G.M., Carrozza, C., Bianchi, A. and Doglietto, F. (2010) Two diagnostic pitfalls mimicking a prolactin-secreting microadenoma. *J. Clin. Endocrinol. Metab.*, 95: 5171.
16. Vallette-Kasic, S., Morange-Ramos, I., Selim, A., Gunz, G., Morange, S., Enjalbert, A., et al. (2002) Macroprolactinemia revisited: A study on 106 patients. *J. Clin. Endocrinol. Metab.*, 87: 581–88.

17. Asa, S.L., Horvath, E. and Kovacs, K. (1993) Pituitary neoplasms: An overview of the clinical presentation, diagnosis, treatment and pathology. In *Endocrine Tumors* (E.L. Mazzaferri and N.A. Samaan, eds.), pp. 77–112. Boston, MA: Blackwell Scientific Publications.
18. Torre, D.L. and Falorni, A. (2007) Pharmacological causes of hyperprolactinemia. *Ther. Clin. Risk Manag.,* 3: 929–51.
19. Gillam, M.P., Molitch, M.E., Lombardi, G. and Colao, A. (2006) Advances in the treatment of prolactinomas. *Endocr. Rev.,* 27: 485–534.
20. Martin, N.M., Tan, T. and Meeran, K. (2009) Dopamine agonists and hyperprolactinaemia. *BMJ,* 338: b381.
21. Fleseriu, M., Lee, M., Pineyro, M.M., Skugor, M., Reddy, S.K., Siraj, E.S., et al. (2006) Giant invasive pituitary prolactinoma with falsely low serum prolactin: The significance of "hook effect". *J. Neurooncol.,* 79: 41–3.
22. Fernandez, A., Karavitaki, N. and Wass, J.A. (2010) Prevalence of pituitary adenomas: A community-based, cross-sectional study in Banbury (Oxfordshire, UK). *Clin. Endocrinol. (Oxf.),* 72: 377–82.
23. Arafah, B.M. and Nasrallah, M.P. (2001) Pituitary tumors: Pathophysiology, clinical manifestations and management. *Endocr. Relat. Cancer,* 8: 287–305.
24. Cushing, H. (1932) The basophil adenomas of the pituitary body and their clinical manifestations (pituitary basophilism). *Bull. Johns Hopkins Hosp.,* 50: 137–95.
25. de Bruin, C., Pereira, A.M., Feelders, R.A., Romijn, J.A., Roelfsema, F., Sprij-Mooij, D.M., et al. (2009) Coexpression of dopamine and somatostatin receptor subtypes in corticotroph adenomas. *J. Clin. Endocrinol. Metab.,* 94: 1118–24.
26. Sanno, N., Teramoto, A., Osamura, R.Y., Horvath, E., Kovacs, K., Lloyd, R.V., et al. (2003) Pathology of pituitary tumors. *Neurosurg. Clin. N. Am.,* 14: 25–39.
27. Vezina, J.L., Hardy, J. and Yamashita, M. (1975) Microadenomas and hypersecreting pituitary adenomas. *Arq. Neuropsiquiatr.,* 33: 119–27. [article in Portuguese].
28. Knosp, E., Steiner, E., Kitz, K. and Matula, C. (1993) Pituitary adenomas with invasion of the cavernous sinus space: A magnetic resonance imaging classification compared with surgical findings. *Neurosurgery,* 33: 610–17.
29. Chacko, G., Chacko, A.G., Lombardero, M., Mani, S., Seshadri, M.S., Kovacs, K., et al. (2009) Clinicopathologic correlates of giant pituitary adenomas. *J. Clin. Neurosci.,* 16: 660–65.
30. Bailey, P. (1898) Pathological report of a case of acromegaly, with special reference to the lesions in the hypophysis cerebri and in the thyroid gland; and a case of haemorrhage in to the pituitary. *Philadelphia Med. J.,* 1: 789–92.
31. Wakai, S., Fukushima, T., Teramoto, A. and Sano, K. (1981) Pituitary apoplexy: Its incidence and clinical significance. *J. Neurosurg.,* 55: 187–93.
32. Rajasekaran, S., Vanderpump, M., Baldeweg, S., Drake, W., Reddy, N., Lanyon, M., et al. (2011) UK guidelines for the management of pituitary apoplexy. *Clin. Endocrinol. (Oxf.),* 74: 9–20.
33. Semple, P.L., Jane, J.A., Jr. and Laws, E.R., Jr. (2007) Clinical relevance of precipitating factors in pituitary apoplexy. *Neurosurgery,* 61: 956–61.
34. Randeva, H.S., Schoebel, J., Byrne, J., Esiri, M., Adams, C.B. and Wass, J.A. (1999) Classical pituitary apoplexy: Clinical features, management, and outcome. *Clin. Endocrinol. (Oxf.),* 51: 181–88.
35. Bills, D.C., Meyer, F.B., Laws, E.R., Jr., Davis, D.H., Ebersold, M.J., Scheithauer, B.W., et al. (1993) A retrospective analysis of pituitary apoplexy. *Neurosurgery,* 33: 602–08.
36. Ayuk, J., McGregor, E.J., Mitchell, R.D. and Gittoes, N.J. (2004) Acute management of pituitary apoplexy—surgery or conservative management? *Clin. Endocrinol. (Oxf.),* 61: 747–52.

37. Sibal, L., Ball, S.G., Connolly, V., James, R.A., Kane, P., Kelly, W.F., et al. (2004) Pituitary apoplexy: A review of clinical presentation, management, and outcome in 45 cases. *Pituitary,* 7: 157–63.
38. Leyer, C., Castinetti, F., Morange, I., Gueydan, M., Oliver, C., Conte-Devolx, B., et al. (2010) A conservative management is preferable in milder forms of pituitary tumor apoplexy. *J. Endocrinol. Invest.,* 33.
39. Gruber, A., Clayton, J., Kumar, S., Robertson, I., Howlett, T.A. and Mansell, P. (2006) Pituitary apoplexy: Retrospective review of 30 patients—is surgical intervention always necessary? *Br. J. Neurosurg.,* 20: 379–85.
40. Maccagnan, P., Macedo, C.L., Kayath, M.J., Nogueira, R.G. and Abucham, J. (1995) Conservative management of pituitary apoplexy: A prospective study. *J. Clin. Endocrinol. Metab.,* 80: 2190–97.
41. Zayour, D.H., Selman, W.R. and Arafah, B.M. (2004) Extreme elevation of intrasellar pressure in patients with pituitary tumor apoplexy: Relation to pituitary function. *J. Clin. Endocrinol. Metab.,* 89: 5649–54.
42. Turgut, M., Ozsunar, Y., Basak, S., Guney, E., Kir, E. and Meteoglu, I. (2010) Pituitary apoplexy: An overview of 186 cases published during the last century. *Acta Neurochir. (Wien.),* 152: 749–61.
43. Fraser, L.A., Lee, D., Cooper, P. and Van Uum, S. (2009) Remission of acromegaly after pituitary apoplexy: Case report and review of literature. *Endocr. Pract.,* 15: 725–31.
44. Gheorghiu, M.L., Baculescu, N., Dumitrascu, A., Purice, M. and Coculescu, M. (2006) Acromegaly remission after spontaneous pituitary adenoma apoplexy. *Acta Endocrinol. (Buc.),* 2: 495. [abstract].
45. Gheorghiu, M.L., Niculescu, D., Dumitrascu, A. and Coculescu, M. (2008) Pituitary stone in long-standing acromegaly with spontaneous remission. *Acta Endocrinol. (Buc.),* 4: 203–10.
46. Kamiya, Y., Jin-No, Y., Tomita, K., Suzuki, T., Ban, K., Sugiyama, N., et al. (2000) Recurrence of Cushing's disease after long-term remission due to pituitary apoplexy. *Endocr. J.,* 47: 793–97.
47. Randall, B.R. and Couldwell, W.T. (2010) Apoplexy in pituitary microadenomas. *Acta Neurochir. (Wien.)* 152: 1737–40.

2 Immunohistochemistry and Electron Microscopy as Evaluation Criteria in Tumor Classification in the Deciphering of Pituitary Adenomas

Pituitary adenomas are benign neoplasms originating in adenohypophyseal cells representing 15% of all intracranial tumors, which express a wide hormonal and proliferation behavior. Pathology data are the gold standard of diagnosis and classification and give important information for therapy. When pituitary adenomas are hormonally inactive, they are not clinically detected and are discovered only incidentally (by CT/MRI scanning) or when they grow enough to determine symptoms of compression on normal pituitary or optic pathway.

Pituitary adenomas with normal plasma hormone levels, considered "nonsecreting" adenomas, do not always correspond with pathology. Plurihormonal pituitary adenomas have raised many controversies over time, beginning with their discovery, which contradicted the traditional dogma of "one cell, one hormone." From the introduction of immunohistochemical staining in the usual algorithm of diagnosis of pituitary adenomas, evidences accumulated that the adenohypophyseal cells did not have a straight specialization, and the group of hormonally silent adenomas was not entirely inactive as had been believed.

In the past, classification of pituitary adenomas was based on their histologic staining characteristics: acidophilic, basophilic, and chromophobic types. Today, we know that acidophilic adenomas can synthesize PRL or can be hormonally inactive and chromophobe adenomas can secrete in excess GH, PRL, ACTH, and glycoprotein hormones (FSH, LH, and TSH).

Today, due to modern methods of cellular study (such as immunohistochemistry, EM, and *in situ* hybridization), tinctorial classification is of no use because of the failure to bring information about hormone content and biological activity of the tumor cells. For correlation between hormone production, secretory activity and cytogenesis, immunohistochemistry, and transmission, EM investigation is needed.

The World Health Organization (WHO) classification of pituitary adenomas is based on morphological features and hormone production and secretion, also taking into consideration findings from clinical symptoms to imaging procedures (Box 2.1).

Prolactinomas

These tumors are chromophobic and contain spheroid nuclei with prominent nucleolus. Psammoma bodies and interstitial amyloid deposits, which are seen occasionally,

Molecular Pathology of Pituitary Adenomas. DOI: 10.1016/B978-0-12-415830-6.00002-0

Box 2.1 Pathology Classification of Pituitary Adenomas

- GH-producing adenoma
  - Densely granulated adenoma
  - Sparsely granulated adenoma
  - Mixed adenoma
  - Mammosomatotroph adenoma
  - Acidophilic stem cell adenoma
- PRL-producing adenoma
  - Densely granulated adenoma
  - Sparsely granulated adenoma
  - Acidophilic stem cell adenoma
- TSH-producing adenoma
- ACTH-producing adenoma
  - Silent ACTH cell adenoma (subtypes 1 and 2)
- Gonadotrophin-producing adenoma
- Unclassified adenoma
  - Unusual plurihormonal adenoma
  - Null cell adenoma/negative immunostaining
  - Oncocytoma
- Atypical adenoma
- Carcinoma
- Others

are typical for this adenoma type. The dotlike paranuclear localization of PRL immunoreactivity, the so-called Golgi pattern, is a characteristic diagnostic feature for PRL production. Misplaced exocytosis, representing granule extrusion into the extracellular space, is considered an electron microscopic hallmark for PRL secretion. Most prolactinomas are sparsely granulated, with secretory vesicles of about 150–500 nm, while a few of them are densely granulated, with big secretory granules of 400–1000 nm.

GH-Producing Adenomas

Densely Granulated Somatotroph Adenomas

Densely granulated somatotroph adenomas correspond to classic acidophil adenomas. The tumor cells are similar to nontumorous somatotroph cells. They are large polyhedrals with round and often prominent nuclei. The cytoplasm contains abundant, large secretory granules and demonstrates strong and diffuse immunoreactivity for GH. Up to 50% of these tumors show plurihormonal immunophenotype and are also positive for α- and/or β-subunits of TSH, FSH, or LH [1]. At EM, the average storage vesicles are about 400–600 nm.

Sparsely Granulated Somatotroph Adenomas

Sparsely granulated somatotroph adenomas consist of chromophobic cells lacking similarities to any cell type of the nontumorous adenohypophyseal gland. The adenoma cells are smaller, with conspicuous nucleoli. The diagnostic feature of this variant is the presence of pale acidophilic, spheroid cytoplasmic inclusions known as *fibrous bodies*. Adenoma cells containing fibrous bodies show peripheral displacement of the nuclei, often with crescent formation, and may harbor multinucleated or pleomorphic nuclei. The cytoplasm contains sparse and small secretory granules (100–250 nm), showing variable immunoreactivity for GH. The fibrous bodies consist of concentric aggregates of keratin-intermediate filaments that are strongly reactive for low-molecular-weight cytokeratins, particularly for keratin 8 [2].

Mixed Somatotroph–Lactotroph Adenomas

Mixed somatotroph–lactotroph adenomas are uncommon tumors that are bimorphous, consisting of somatotroph and lactotroph cells. The two different adenoma cell populations are variably admixed, and each particular cell type shows separate immunoreactivity for GH and PRL. They arise from a common progenitor cell which entitles them with an increased invasivity.

Mammosomatotroph Adenomas

Mammosomatotroph adenomas are monomorphous tumors consisting of a single cell type. By histology, the adenoma cells are acidophilic, identical to densely granulated somatotroph adenomas; and by immunohistochemistry, GH and PRL reactivities are colocalized in the same cells. By EM, the cells resemble densely granulated somatotroph adenomas of the nontumorous pituitary, with big secretory granules of up to 500–1500 nm. In addition, they show misplaced exocytosis, a characteristic feature of PRL-secreting cells. Some cells may contain small fibrous bodies [3].

Acidophil Stem Cell Adenomas

Acidophil stem cell adenomas are unusual, monomorphous, slightly acidophilic tumors. They show nuclear pleomorphism with coarse chromatin and prominent nucleoli. The large cytoplasmic vacuoles corresponding to giant mitochondria, the presence of fibrous bodies immunoreactive for keratin 8, and the misplaced exocytosis represent the key diagnostic features of this adenoma type. Immunoreactivity for PRL is often associated with less-pronounced positivity for GH. However, only EM can establish this diagnosis, showing small secretory vesicles of 50–250 nm.

ACTH-Producing Adenomas

The histology of ACTH-producing adenomas shows typical basophilic adenomas consisting of monomorphous round cells with plump cytoplasm and ovoid nuclei with conspicuous nucleoli. The adenoma cells are often strongly periodic acid Schiff (PAS) PAS-positive, although adenoma cells containing abundant keratin-intermediate filaments may be weakly PAS-positive. Some aggressive tumors (developing Nelson syndrome) may show nuclear pleomorphism and apoptotic figures. Crooke's hyaline change, corresponding to accumulated cytokeratin-intermediate filaments in the cytoplasm of nonneoplastic corticotrophs, is commonly present. By immunohistochemistry, Pro-opiomelanocortin (POMC) derivatives ACTH and Melanocyto Stimulating Hormone (MSH) are positive. In some cases, ACTH reactivity is less pronounced due to the presence of abundant cytokeratin filaments. Silent ACTH adenomas are defined by lack of Cushing clinical signs of hypercortisolism, and are usually more aggressive. Crooke's adenomas represent ACTH-producing tumors with massive hyaline deposits in adenoma cells. By EM, depending on the size (250–700 nm) and number of the secretory granules, corticotroph adenomas can be classified as either densely granulated or sparsely granulated.

Thyrotrophin-Producing Adenomas

Thyrotrophin-producing adenomas are very rare, and they induce a clinical syndrome of central hyperthyroidism, characterized by lack of negative feedback of excess thyroid hormones upon TSH (i.e., inadequate TSH secretion). By histology, the chromophobic adenoma cells are polyhedral or polar and show varying degrees of nuclear pleomorphism. The PAS stain reveals positive tiny cytoplasmic granules corresponding to lysosomes. Immunohistochemistry is positive for α- and β-TSH subunits. Immunohistochemistry and EM highlight the elongated cytoplasmic processes of the tumor cells. The small secretory granules (50–250 nm) are arranged along the cytoplasmic membrane and accumulate in the cell process.

Gonadotrophin-Producing Adenomas

Typically, gonadotrophin-producing adenomas consist of round, small to mid-sized polar cells, often with a perisinusoidal pattern with elongated cytoplasmic processes forming characteristic pseudorosettes or pseudopapillae.

They show uniform, spherical to slightly ovoid nuclei with scant nucleoli. The cytoplasm is often granular and acidophilic due to oncocytic transformation. The PAS stain is negative, although some adenoma cells may show PAS-positive granules corresponding to lysosomes.

The typical immunophenotype includes one or more of β-FSH, β-LH, and α-glycoprotein hormone subunits. The distribution of immunoreactivity is heterogeneous and mostly focal, and immunoreactivities for β-FSH and β-LH are not necessarily confined to the same cells. Immunohistochemistry demonstrates the cytoplasmic configuration and highlights the cytoplasmic processes where the secretory granules accumulate. Gonadotrophin-producing adenomas are also immunoreactive for synaptophysin and chromogranin A, and for inhibin and activin subunits and the transcription factor steroidogenic factor 1 (SF1).

EM shows cells with polar cells and reveals tiny secretory granules (50–200 nm) lining up along cytoplasmic membranes, in addition to those accumulating in the cytoplasmic processes. This feature is also known as "glycoprotein hormone differentiation." In females, the Golgi apparatus shows characteristic vesicular, "honeycomb-like" transformation. This unique abnormality represents a diagnostic feature, particularly in tumors with inconclusive or negative immunoreactivity for gonadotrophin hormones. Gonadotroph adenomas often show high accumulation of mitochondria (over 40% of cytoplasm volume) in the cytoplasm, which is known as *oncocytic transformation*.

Null Cell Adenomas

By histology, null cell adenomas are typically chromophobic, with a pseudopapillary pattern of growth that often features pseudorosette formation. They contain round, focally conspicuous nucleoli and no marked nuclear pleomorphism or mitotic activity. Despite the weak immunoreactivity for gonadotrophins, sensitive molecular techniques reveal their presence. In addition, these tumors are immunoreactive for chromogranin A and synaptophysin. EM demonstrates poorly developed cytoplasmic membranous organelles, including scattered arrays of rough endoplasmic reticulum and Golgi apparatus saccules. The secretory granules are sparse and small, measuring 100–250 nm. An increased number and higher-volume density of cytoplasmic mitochondria, leading to oncocytic transformation, are also frequent.

It seems that null cell adenomas and oncocytomas belong to the same entity with gonadotroph adenomas. However, other nonfunctioning adenomas, such as silent GH [4], silent ACTH-producing adenomas [5], and silent subtype 3 adenomas [6], should be differentiated because they may show a higher proliferation rate and recur more frequently than null cell adenomas.

False-Positive and -Negative Results on Immunohistochemistry

Chemical fixation is critically important because it changes the chemical and physical makeup of biological tissue and can render tissue antigens unrecognizable to their antisera for immunoreaction. In glutaraldehyde-fixed tissue, the antigen sensitivity is inversely proportional to the concentration of glutaraldehyde and the time for which it is used. Formaldehyde penetrates faster and does not cross-link proteins like glutaraldehyde. For immersion–fixation, if the size of the tissue samples is small (1–3 mm^3) in at least one dimension and the fixative used is based on neutral-buffered formalin, it is possible to achieve reasonable levels of fixations while preserving tissue reactivity [7]. For the EM-immunogold method, a mixture of 4% of formaldehyde and 0.2% glutaraldehyde for 2–4 h is optimal for fixation and antigen preservation.

For immunocytochemistry, buffered formaldehyde 4% offers good tissue preservation, but the fixation time should not last longer than 12 h. In the interpretation of immunoreactivity at the LM level, we must be cautious in evaluating positive cells in a fragmented tissue excised by the surgeon because many times, the limits between normal and tumor tissue are not very clear. If the normal tissue is not identified from surgery samples using basic or special staining, it could be misinterpreted as a false-positive reaction of a plurihormonal adenoma.

The characteristic histological feature on hematoxylin-eosin stain of pituitary adenoma includes sheets or a cordonal architecture of polygonal monomorphous cells, with regular or moderate pleomorph nuclei. The mitotic activity is usually low. The cellular cytoplasm could be acidophilic, basophilic, or chromophobic, depending on the type and the amount of secretory content in the cell. The reticulin network is destroyed, with most of these tumors having a soft, gelatinous consistency. This cellular monomorphism and the absence of a reticulin network are the difference between a pituitary adenoma and nonneoplastic tissue from the anterior pituitary. The polymorphous pituitary adenomas need a Gomori stain in order to differentiate the normal tissue with preserved reticulin network. The functional status of adenoma could not be predicted on a histological basis.

After we eliminate the normal tissue, it is still possible to obtain a false-positive immunoreaction from endogenous peroxidase. In addition, an intratumoral hemorrhage or necrosis can lead to high levels of endogenous peroxidase. Here, the counterstaining with hematoxylin and the experience of the examiner are crucial in tissue evaluation. False-negative reactions on immunohistochemistry could be eliminated easily using a positive control in the same protocol from normal pituitary autopsied tissue.

Apart from technical problems related to antibody activity/specificity, secondary antibodies and reporting enzymes, rough fixation or paraffin embedding at high temperature can affect the immunoreaction of sensitive antigens. For sensitive antigens, procedures for antigen retrieval like microwave or boiling can be tried. At the EM level, polymorphous adenomas should be differentiated from normal cells by their tumor characteristics.

Discordances Between Serum Biochemistry and Immunohistochemistry

Even if the cellular hormonal immunoreactivity is common in pituitary tumors, this is not always correlated with the synthesis or release rate of hormones. Systematic immunohistochemical evaluation shows that 20–50% of pituitary tumors are plurihormonal. Commonly, there are tumors with GH, PRL, or TSH production. The possible explanation could be the same transcription factor Pit-1, which regulates gene expression for GH, PRL, and TSH. Other plurihormonal tumors also could be associated with GH, FSH, PRL, or TSH [8,9].

A clinically nonfunctioning GH immunoreactive adenoma could hide a silent somatotroph adenoma [4], in fact a sparsely granulated adenoma. There are pathology proofs and cell culture studies in favor of silent lactotroph adenomas [10], thyrotroph adenomas [6], and silent corticotroph adenomas [5], which are not associated with excessive autonomous hormonal secretion, but they have specific immunocytochemical and ultrastructural cell features. The frequent silent adenomas are gonadotroph adenomas, which are more classified morphologically than clinically diagnosed. These adenomas present immunoreactivity for α-subunit (SU), β-FSH, and β-LH [9,11]. In our study, 25/92 (26.3%) tumors initially diagnosed as clinically nonfunctioning adenomas were tissue that was immunoreactive for FSH and/or LH [12].

Immunocytochemical studies [13] showed *in vitro* that most nonsecreting pituitary adenomas are formed by cells that contain intact glycopeptidic hormones and/or free α- or β-subunits. Approximately 86% of clinically nonfunctioning adenomas have a positive stain for at least one glycopeptidic hormone. An mRNA analysis proved biosynthesis of subunits of glycopeptidic hormone in most pituitary tumors [14]. Other authors have written about GH synthesis and secretion in clinically nonfunctioning adenomas [4].

Clinically "mute" adenomas have no clinical expression of produced hormones, because they either produce inactive molecules or synthesize but do not release hormones from cells in a sufficient amount to raise the blood level and to determine a clinical effect. Other hypotheses include a lysosomal catabolism of the hormone inside cells or an alteration of posttranslational processing and release of the hormone.

Discordances Between LM and EM

The recent proposals for WHO classification of pituitary adenomas [15] are based on a synthesis of pathological, immunocytochemical, and ultrastructural data. The fact that not all positive tumors at the optic level are confirmed at electronic level can be derived from different methods of antigen preservation because of the small dimensions of the tissue examined at EM (around 1 mm^3).

Therefore, it is necessary that many tissue fragments from different parts of the tumor be embedded for EM. Also very important is the time between the moment of surgical removal and the embedding, which should not be more than 4 h in order to better preserve the antigens. Another aspect could be related to the number of secretory granules and the fact that sparsely granulated adenomas (GH, PRL, and ACTH) have an absent or low immunoreactivity [16] at the LM level.

Pituitary nonfunctioning adenomas are divided into two groups: null cell adenomas and pituitary oncocytomas. The EM feature reflects a very low hormonal activity of these tumors. Often, cases with oncocytoma features on EM are invasive tumors with SSE. They present immunoreactivity for gonadotrophs, suggesting a differential diagnosis between a gonadotroph adenoma with marked oncocytic transformation and a functional oncocytoma. The presence of numerous small secretory granules in tumor cells is a sign of tumor activity. The immunoreactivity on the EM level could not be determined.

Oncocytomas appear to be the variant of null cell adenomas that presents marked oncocytic transformation. A variable number of oncocytic cells also could be observed in adult normal pituitary tissue [17,18] and in a group of pituitary tumors, most frequently in somatotropinomas [19]. The number of oncocytic cells is proportional to the tumor recurrence and appear most frequently in GH-secreting adenomas with low GH serum level and high tumor volume. The oncocytic transformation of pituitary tumors indicates a low differentiation and/or aggressive behavior, with a decrease of endocrine function of the tumor [20].

The genes of glycoproteic hormones are expressed in clinically nonfunctioning pituitary adenomas [21]. There are many similarities between gonadotroph adenomas and null cell adenomas. Null adenomas synthesize glycoproteic hormones detected in cell culture and by *in situ* hybridization. It was hypothesized that gonadotroph hormones from null cell adenomas are not reflecting a gonadotroph origin directly, but rather represent a predominant pluripotent cell differentiation toward a gonadotroph cell line. Future studies will probably bring new evidence about cell origins.

References

1. Kontogeorgos, G. (2005) Classification and pathology of pituitary tumors. *Endocrine*, 28: 27–35.
2. Obari, A., Sano, T., Ohyama, K., Kudo, E., Qian, Z.R., Yoneda, A., et al. (2008) Clinicopathological features of growth hormone-producing pituitary adenomas: Difference among various types defined by cytokeratin distribution pattern including a transitional form. *Endocr. Pathol.*, 19: 82–91.
3. Horvath, E. and Kovacs, K. (1978) Morphogenesis and significance of fibrous bodies in human pituitary adenomas. *Virchows Arch. B Cell Pathol.*, 27: 69–78.
4. Wade, A.N., Baccon, J., Grady, M.S., Judy, K.D., O'Rourke, D.M. and Snyder, P.J. (2011) Clinically silent somatotroph adenomas are common. *Eur. J. Endocrinol.*, 165: 39–44.
5. Psaras, T., Honegger, J., Buslei, R., Saeger, W., Klein, D., Capper, D., et al. (2007) Atypical type II silent corticotrophic adenoma developing into Cushing's disease upon second recurrence. *Exp. Clin. Endocrinol. Diabetes*, 115: 610–15.
6. Erickson, D., Scheithauer, B., Atkinson, J., Horvath, E., Kovacs, K., Lloyd, R.V., et al. (2009) Silent subtype 3 pituitary adenoma: A clinicopathologic analysis of the Mayo Clinic experience. *Clin. Endocrinol. (Oxf.)*, 71: 92–99.
7. Fazekas, I., Hegedus, B., Bacsy, E., Kerekes, E., Slowik, F., Balint, K., et al. (2005) Characterization of human pituitary adenomas in cell cultures by light and electron microscopic morphology and immunolabeling. *Folia Histochem. Cytobiol.*, 43: 81–90.
8. Coculescu, M., Badiu, C., Galoiu, S., Caragheorgheopol, A., Stancu, C. and Morris, J. (2006) Evolution under complex therapy of acromegaly due to a pituitary plurihormonal adenoma with colocalisation of GH and FSH. *Acta Endocrinol. (Buc.)*, 2: 337–48.
9. Al-Sharim, M., Scheithauer, B.W., Horvath, E., Kovacs, K., Smyth, H., Coire, C., et al. (2009) Plurihormonal gonadotroph cell pituitary adenoma: Report of a unique case. *Clin. Neuropathol.*, 28: 182–87.
10. Kageyama, K., Ikeda, H., Nigawara, T., Sakihara, S. and Suda, T. (2007) Expression of adrenocorticotropic hormone, prolactin and transcriptional factors in clinically nonfunctioning pituitary adenoma. *Endocr. J.*, 54: 961–68.

11. Chacko, G., Chacko, A.G., Lombardero, M., Mani, S., Seshadri, M.S., Kovacs, K., et al. (2009) Clinicopathologic correlates of giant pituitary adenomas. *J. Clin. Neurosci.*, 16: 660–65.
12. Badiu, C., Dumbrava, M., Stancu, C., Ciubotaru, V. and Coculescu, M. (2006) Immunoreactivity for glycoproteic hormones and tumor size in pituitary adenomas. *Acta Endocrinol. (Buc.)*, 2: 1–9.
13. Saccomanno, K., Spada, A., Bassetti, M., Gil-del-Alamo, P. and Faglia, G. (1994) Immunodetection of chorionic gonadotrophin and its subunits in human nonfunctioning pituitary adenomas. *J. Clin. Endocrinol. Metab.*, 78: 1103–07.
14. Katznelson, L., Alexander, J.M., Bikkal, H.A., Jameson, J.L., Hsu, D.W. and Klibanski, A. (1992) Imbalanced follicle-stimulating hormone beta-subunit hormone biosynthesis in human pituitary adenomas. *J. Clin. Endocrinol. Metab.*, 74: 1343–51.
15. Al-Shraim, M. and Asa, S.L. (2006) The 2004 World Health Organization classification of pituitary tumors: What is new? *Acta Neuropathol.*, 111: 1–7.
16. Naritaka, H., Kameya, T., Sato, Y., Furuhata, S., Otani, M. and Kawase, T. (1999) Morphological characterization and subtyping of silent somatotroph adenomas. *Pituitary*, 1: 233–41.
17. Diehl, S., Weber, T., Hamester, U., Saeger, W. and Ludecke, D.K. (1988) Light microscopical morphometric classification of oncocytic and chromophobe pituitary adenomas. *Exp. Clin. Endocrinol.*, 91: 341–49.
18. Nishioka, H., Ito, H., Hirano, A. and Kato, Y. (1999) Immunocytochemical demonstration of oncocytes in normal adenohypophysis. *Acta Neuropathol.*, 97: 40–44.
19. Nishioka, H., Ito, H. and Hirano, A. (1999) Immunohistochemical demonstration of oncocytes in nongonadotrophic pituitary adenomas. *Virchows Arch.*, 435: 428–33.
20. Niveiro, M., Aranda, F.I., Paya, A., Boix, E., Peiro, G. and Pico, A. (2004) Oncocytic transformation in pituitary adenomas: Immunohistochemical analyses of 65 cases. *Arch. Pathol. Lab. Med.*, 128: 776–80.
21. Desai, B., Burrin, J.M., Nott, C.A., Geddes, J.F., Lamb, E.J., Aylwin, S.J., et al. (1995) Glycoprotein hormone alpha-subunit production and plurihormonality in human corticotroph tumours—an *in vitro* and immunohistochemical study. *Eur. J. Endocrinol.*, 133: 25–32.

3 Proliferation

Deregulation in proliferation control may be regarded as a major driver in tumorigenesis. Most of the molecular checkpoints in cell cycle control are overridden in tumorigenesis, so a brief review of them will point out the major actors to be monitored.

The normal cell cycle can be influenced, stimulated, or inhibited by many different factors, which may also be important for the tumor cell cycle.

In the G1 phase, there are two types of DNA damage responses: p53-dependent and p53-independent pathways. The p53-dependent responses inhibit cyclin-dependent kinases (CDKs) through the upregulation of genes encoding cyclin-dependent kinases inhibitors (CDKIs) mediated by the p53 protein, whereas the p53-independent mechanisms inhibit CDKs through the inhibitory phosphorylation of Cdk2. Failure of the DNA damage checkpoints in G1 leads to mutagenic replication of damaged templates and other replication defects.

The arrest at the G1/S checkpoint is mediated by the action of p53, a widely known tumor suppressor protein. By the loss of p53 functions as a result of mutations in cancer, cells skip the G1/S checkpoint [1]. The p53 protein is rapidly induced in response to damaged DNA. A number of kinases, phosphatases, histone acetylases, and ubiquitin-conjugating enzymes regulate the stability as well as the transcriptional activity of p53 after DNA damage.

G2/M checkpoints include the checks for damaged DNA and unreplicated DNA, as well as checks that ensure that the genome is replicated once, and only once, per cell cycle. If cells pass these checkpoints, they follow the normal transition to the M phase. However, if any of these checkpoints fails, mitotic entry is prevented by specific G2/M checkpoint events.

The G2/M checkpoints can fail due to the presence of unreplicated DNA or damaged DNA. In such instances, the cyclin-dependent kinase Cd2 (Cdk1) is maintained in its inactive, phosphorylated state, and mitotic entry is prevented.

In the late G1 phase, after transcription and translation had occurred, and in the G2 phase, cell cycle checkpoints are installed for stopping the replication process that is mediated by protein kinases (which activate enzymes and other proteins by phosphorylation) and controlled by proteins called *cyclins.* In the G1 phase, cyclins D1, D2, and D3 and the protein kinases 2, 4, 5, and 6 are regulatory. Cyclin-dependent protein kinases (CDKs) are inhibited by multiple inhibitory proteins, especially p15, p16, p18, p19, p21, and p27. For G1-S transition, cyclin E and Cdk2 regulated by the retinoblastoma gene are necessary; they are inhibited by the p53 protein, which enhances the gene transcription of p21.

Proliferation markers are widely used in general pathology and pituitary pathology. They should enable differentiation of aggressive or rapidly growing tumors from slow-growing ones, as cellular atypia is not helpful for identifying aggressive

Molecular Pathology of Pituitary Adenomas. DOI: 10.1016/B978-0-12-415830-6.00003-2

adenomas of the pituitary. Prognosis is based on the mitotic index. Several biomarkers can be evaluated by appropriate immunodetection techniques: cyclins A, B, D, and E, proliferating cell nuclear antigen (PCNA) Ki-67, proliferating inhibitory proteins p16, p27, and p53, and DNA topoisomerase IIα. A marker for apoptosis and its inhibitors also may be important. According to Kontogeorgos and Kovacs [2], the MIB-1 LI is the most reliable marker.

Cyclins control transitions between phases of the cell cycle by activating CDKs. Cyclins D1 and D3 allow the cell to proceed into the S phase. Cyclins A, B, and E were demonstrated in all adenoma types and were significantly higher in macroadenomas compared to microadenomas [3].

Gene expression of cyclins D1, E1, and B1 is significantly suppressed in ACTH-secreting microadenomas, compared to nonfunctioning adenomas [4]. Cyclin D1 gene expression positively correlates with the B1 and E1 cyclins. Positive immunostaining for p16 and negative immunostaining for cyclin D1 were more frequent in secretory adenomas than in nonsecretory adenomas (NFA, non-functional adenomas); there were positive correlations between mRNA and protein expressions of p16 and cyclin D1. Thus, it is suggested that upregulated CDKN2A with the concomitant downregulated cyclin gene family is partly involved in the small size of ACTH-secreting adenoma.

Cyclin D1 was immunohistologically demonstrated more frequently in nonfunctioning and aggressive adenomas than in other adenomas [5]. The amplification of cyclin D1 gene was shown by Yu and Melmed [6]. Turner et al. [3] found cyclin D1 expression in pituitary adenomas related to size and tumor regrowth. The differences between recurrent and nonrecurrent tumors were related to reduced bcl-2 expression, increased cell proliferation, more cells in the G2/M stage, and reduced cell differentiation with more aggressive subsequent behavior.

Cyclin A demonstrated a positive linear correlation with the Mib-index [7]; in recurrent adenomas, the LI was more than double in comparison with nonrecurrent adenomas. The PCNA (cyclin-proliferating cell nuclear antigen) index is very low in normal pituitary, while in adenomatous pituitary, higher values were detected [8].

The Ki-67 antigen, a marker of cellular proliferation, has been studied extensively in pituitary neoplasia. It is relevant for various clinicopathological parameters, such as tumor subtype, size, invasiveness, and recurrence, as well as patient age and sex. Generally, pituitary tumors behaving aggressively have increased Ki-67 LIs. Whereas a number of studies found conclusive associations of Ki-67 LIs with aggressive behavior, size, and/or adenoma subtype, others fail to do so.

The Ki-67 protein expression is a nuclear feature during cell proliferation. Its expression is detected by the monoclonal antibody MIB-1, and it is expressed as a percentage of immunopositive nuclei in the form of a Ki-67 LI. Various authors reported Ki-67 LIs in pituitary adenomas at levels ranging from less than 1% to as high as 23% [9].

Tumor size is not necessarily a predictor of the clinical course of pituitary adenomas. Macroadenomas may remain idle or regrow slowly, whereas some microadenomas grow aggressively. Although it may seem intuitive that tumors with increased Ki-67 LIs grow more rapidly and are larger, few studies demonstrated such an association. However, there are studies that did not find a significant difference in Ki-67 labeling of PRL macroadenomas and microadenomas [10].

Fainstein et al. [11] reported a higher Ki-67 index in male patients with prolactinoma compared to female patients ($3.5 \pm 1.2\%$ vs. $1.5 \pm 0.5\%$, $p < 0.0001$). The study showed that macroprolactinomas in men are clinically and biologically more aggressive.

Recurrent adenomas showed a significantly higher Ki-67 LI than nonrecurrent ones [12]. An inverse correlation between age and Mib-1 index was reported by Tanaka et al. [13]: Elderly patients have lower values, and this correlates with a longer time for doubling tumor volume in elderly patients. A positive association between preoperative PRL levels and Mib-1 LI, which was lower in young female patients than in older female and male patients, was demonstrated, while carcinomas and their metastases showed the highest LI [1]. High LIs for Ki-67 and TUNEL (terminal deoxynucleotidyltransferase-mediated dUTP nick end labeling) were associated with recurrent nonfunctioning pituitary adenomas. Tumors with a high level of expression of phospho-Akt, phospho-p44/42 mitogen-activated protein kinase (MAPK), and pituitary tumor transforming gene 1 (PTTG1) were associated with early recurrence; high levels of expression of phospho-CREB (cyclin-Amp (cAMP) responsive-element binding) and ZAC1 were inversely associated with recurrence. In recurrent groups, a higher percentage of Ki-67 positive cells was found, with higher values in the early recurrence group (less than 4 years) compared to the late recurrence group [14].

The p16 gene encodes a physiological inhibitor of the cyclin D–Cdk4 complex and is considered an important tumor suppressor gene. Methylation of the CpG region of the p16 gene is associated with lower expression of p16 protein in pituitary tumors; hypermethylation was found in 71% of nonsecreting adenomas and 29% of gonadotroph adenomas, but not in GH- or PRL-secreting adenomas [15]. PCR (polymerase chain reaction) findings were supported by immunostaining, and hypermethylated tumors were negative for p16 protein nuclear expression. However, invasive and noninvasive adenomas did not show significant differences of expression.

The p27 kip1 (p27) gene encodes an inhibitor of CDK activity. In endocrine and other tumors, p27 nuclear expression is inversely related to the Mib-1 immunostaining.

The expression of the p27 protein is lower in carcinomas than in adenomas [16,17]. A lower mean LI for p27 (47%) was found in recurrent adenomas than in nonrecurrent adenomas (67%) [18].

Generally, p27 expression was inversely related to the proliferation marker Mib-1 [11]. Treatment of PRL adenomas with dopamine agonists did not significantly influence the expression of p27 [19]. Vitamin D3 hypophosphorylates p27 and can accumulate the p27 protein in pituitary adenomas and was found to arrest the growth of ACTH-secreting cells [20].

Point mutations in highly conserved regions of the p53 gene induce conformational alterations that stabilize the protein in transformed cells. In normal cells, the p53 protein is undetectable by immunocytochemistry. Specific monoclonal antibodies for mutant forms of p53 make possible their detection by immunostaining. The loss of p53 function may result from mutations in one or both alleles or from deletions of both alleles. The loss of alleles in the region of chromosome 17p, where the p53 gene is located, frequently occurs in malignant tumors. In many tumors, deletion of one allele and mutation of the other were found [1].

Table 3.1 Distribution of Cases Based on the Ki-67 LI

	Invasive	Noninvasive	Total
IM Ki-67 >3%	22	0	22
IM Ki-67 =3%	7	4	11
IM Ki-67 <3%	5	12	17
Total	34	16	50

Threshold value is 3%.

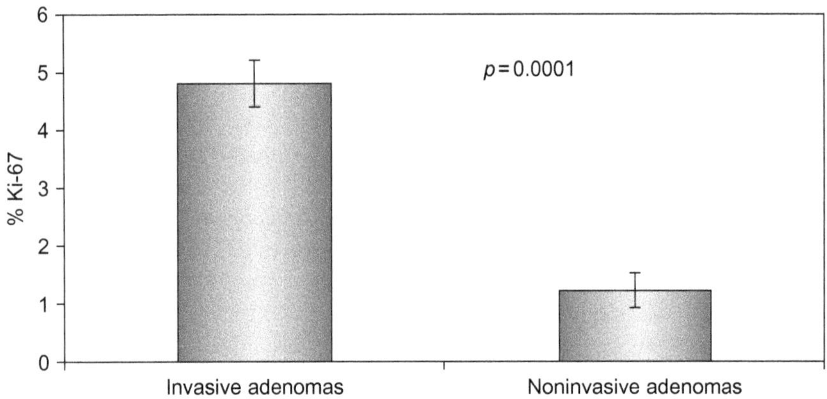

Figure 3.1 The Ki-67 LI in invasive and noninvasive pituitary adenomas.

The expression of p53 has been linked to aggressive tumor behavior. Thapar et al. [21] demonstrated significant association between p53 expression and tumor behavior, with LIs of 0%, 15.2%, and 100% in non-invasive adenomas, invasive adenomas and carcinomas, respectively ($p < 0.001$). Higher p53 expression was found in "aggressive-invasive" tumors compared with those with less-aggressive behavior [22].

The expression of p53 was higher in ACTH carcinomas than in invasive ACTH adenomas (49.9% vs. 37.3%) [23]. Roncaroli et al. detected P53 expression in invasive pituitary adenomas, but not in noninvasive adenomas [24].

According to Kontogeorgos and Kovacs [2], positive p53 protein in pituitary adenomas can be interpreted as a sign of increased or invasive growth, but on the other hand, negative p53 protein does not exclude a more rapid or even malignant tumor growth.

Our investigations revealed a Ki-67 expression in the adenomas group of 3.68%. We noticed a positive correlation between the proliferative activity and tumoral invasiveness (Table 3.1). The mean value for Ki-67 LI was significantly higher in invasive adenomas relative to noninvasive adenomas (Figure 3.1). Some differences also were recorded in functional and nonfunctional adenomas (Figure 3.2). Using an LI of 3% as the threshold discriminating value, the case distribution of Ki-67 expression in invasive and noninvasive adenomas showed that invasive adenomas are correlated

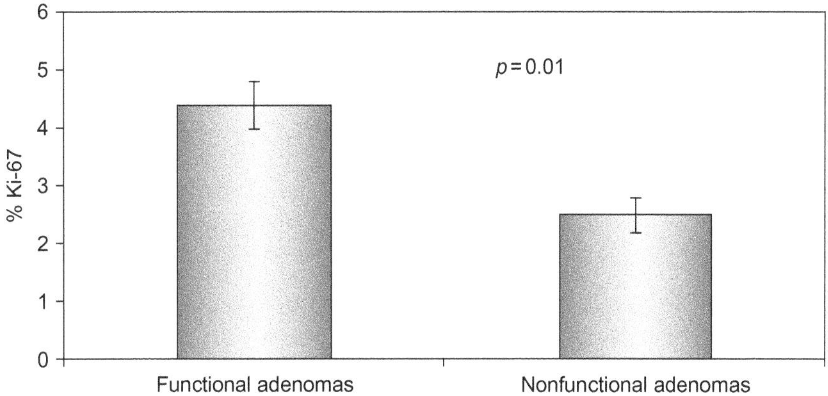

Figure 3.2 The Ki-67 LI in secretory and nonsecretory pituitary adenomas.

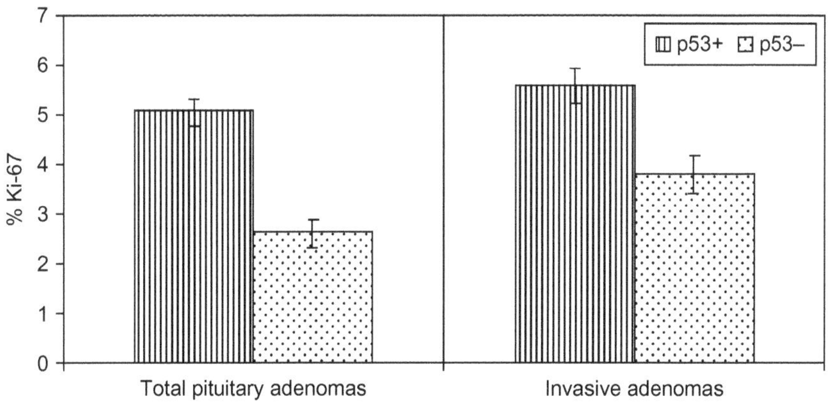

Figure 3.3 Comparison between p53 expression and the Ki-67 LI.

with more than 3% expression (see Table 3.1). We found a positive correlation between the proliferation rates and p53 expression (Figure 3.3), and also between the proliferation rates and apoptotic indices. The results suggest a relationship between tumoral invasiveness and proliferation/apoptotic processes.

References

1. Kuerbitz, S.J., Plunkett, B.S., Walsh, W.V. and Kastan, M.B. (1992) Wild-type p53 is a cell cycle checkpoint determinant following irradiation. *Proc. Natl. Acad. Sci. U.S.A.*, 89(16): 7491–95.
2. Kontogeorgos, G., and Kovacs, K. (eds) (2004) *Molecular Pathology of the Pituitary*. Karger, Basel.

3. Turner, H.E., Nagy, Z., Sullivan, N., Esiri, M.M. and Wass, J.A.H. (2000) Expression analysis of cyclines in pituitary adenomas and the normal pituitary gland. *Clin. Endocrinol. (Oxf.)*, 53: 337–44.
4. Tani, Y., Inoshita, N., Sugiyama, T., Kato, M., Yamada, S., Shichiri, M., et al. (2010) Upregulation of CDKN2A and suppression of cyclin D1 gene expressions in ACTH-secreting pituitary adenomas. *Eur. J. Endocrinol.*, 163(4): 523–29.
5. Jordan, S., Lidhar, K., Korbonits, M., Lowe, D.G. and Grossman, A.B. (2000) Cyclin D and cyclin E expression in normal and adenomatous pituitary. *Eur. J. Endocrinol.*, 143: R1–R6.
6. Yu, R. and Melmed, S. (2010) Pathogenesis of pituitary tumors. *Prog. Brain Res.*, 182: 207–27.
7. Nakabayashi, H., Sunada, I. and Hara, M. (2001) Immunohistochemical analyses of cell cycle–related proteins, apoptosis, and proliferation in pituitary adenomas. *J. Histochem. Cytochem.*, 49: 1193–94.
8. Klencki, M., Kurnatowska, I., Slowinska-Klencka, D., Lewinski, A. and Pawlikowski, M. (2001) Correlation between PCNA expression and AgNOR dots in pituitary adenomas. *Endocr. Pathol.*, 12: 163–69.
9. Prevedello, D.M., Jagannathan, J., Jane, J.A., Lopes, M.B. and Laws, E.R., Jr. (2005) Relevance of high Ki-67 in pituitary adenomas. Case report and review of the literature. *Neurosurg. Focus*, 19: E11.
10. Delgrange, E., Trouillas, J., Maiter, D., Donckier, J. and Tourniaire, J. (1997) Sex-related difference in the growth of prolactinomas: A clinical and proliferation marker study. *J. Clin. Endocrinol. Metab.*, 82: 2102–07.
11. Fainstein, D.P., Glerean, M., Lovazzano, S., Pietrani, M., Christiansen, S., Balzaretti, M., et al. (2010) Gender differences in macroprolactinomas: Study of clinical features, outcome of patients, and ki-67 expression in tumor tissue. *Front. Horm. Res.*, 38: 50–8.
12. Losa, M., Franzin, A., Mangili, F., Terreni, M.R., Barzaghi, R., Veglia, F., et al. (2000) Proliferation index of nonfunctioning pituitary adenomas: Correlations with clinical characteristics and long-term follow-up results. *Neurosurgery*, 47: 1313–18.
13. Tanaka, Y., Hongo, K., Tada, T., Sakai, K., Kakizawa, Y. and Kobayashi, S. (2003) Growth pattern and rate in residual nonfunctioning pituitary adenomas: Correlations among tumour volume doubling time, patient age, and MIB-1 index. *J. Neurosurg.*, 98: 359–65.
14. Noh, T.W., Jeong, H.J., Lee, M.K., Kim, T.S., Kim, S.H. and Lee, E.J. (2009) Predicting recurrence of nonfunctioning pituitary adenomas. *J. Clin. Endocrinol. Metab.*, 94(11): 4406–13.
15. Ruebel, K.H., Jin, L., Zhang, S.Y., Scheithauer, B.W. and Lloyd, R.V. (2001) Inactivation of the p16 gene in human pituitary nonfunctioning tumors by hypermethylation is more common in null cells adenomas. *Endocr. Pathol.*, 12: 281–89.
16. Gaffey, T.A., Scheithauer, B.W., Lloyd, R.V., Burger, P.C., Robbins, P., Fereidooni, F., et al. (2002) Corticotroph carcinoma of the pituitary: A clinicopathological study. *J. Neurosurg.*, 96: 352–60.
17. Korbonits, M., Chahal, H.S., Kaltsas, G., Jordan, S., Urmanova, Y., Khalimova, Z., et al. (2002) Expression of phosphorylated p27(Kip1) protein and Jun activation domain-binding protein 1 in human pituitary tumors. *J. Clin. Endocrinol. Metab.*, 87: 2635–43.
18. Nakabayashi, H., Sunada, I. and Hara, M. (2001) Immunohistochemical analyses of cell cycle–related proteins, apoptosis, and proliferation in pituitary adenomas. *J. Histochem. Cytochem.*, 49: 1193–94.
19. Stefaneanu, L., Kovacs, K., Scheithauer, B.W., Kontogeorgos, G., Reihle, D.L., Sebo, T.J., et al. (2000) Effect of dopamine agonists on lactotroph adenomas of the human pituitary. *Endocr. Pathol.*, 11: 341–52.

20. Liu, W., Asa, S.L. and Ezzat, S. (2002) Vitamin D and its analog EB1089 induce p27 accumulation and diminish association of p27 with Skp2 independent of PTEN in pituitary corticotroph cells. *Brain Pathol.*, 12: 412–19.
21. Thapar, K., Scheithauer, B.W., Kovacs, K., Pernicone, P.J. and Laws, E.R., Jr. (1997) p53 expression in pituitary adenomas and carcinomas: Correlation with invasiveness and tumor growth fractions. *Neurosurgery*, 38(4): 765–70. ; discussion 770–1.
22. Wierinckx, A., Auger, C., Devauchelle, P., Reynaud, A., Chevallier, P., Jan, M., et al. (2007) A diagnostic marker set for invasion, proliferation, and aggressiveness of prolactin pituitary tumors. *Endocr. Relat. Cancer*, 14(3): 887–900.
23. Gaffey, T.A., Scheithauer, B.W., Lloyd, R.V., Burger, P.C., Robbins, P., Fereidooni, F., et al. (2002) Corticotroph carcinoma of the pituitary: A clinicopathological study. *J. Neurosurg.*, 96: 352–60.
24. Roncaroli, F., Scheithauer, B.W., Young, W.F., Horvath, E., Kovacs, K., Kros, J.M., et al. (2003) Silent corticotroph carcinoma of the adenohypophysis—A report of five cases. *Am. J. Surg. Pathol.*, 27: 477–86.

4 Angiogenesis

Introduction

Pituitary adenomas are typically slow-growing and histologically benign tumors, but they can become clinically destructive, invade adjacent structures, and recur after treatment. Unlike many other tumors that become aggressive and appear to depend upon angiogenesis in the process, pituitary adenomas tend to do so through nonangiogenic means [1].

Over the past few decades, much work has been done to better elucidate the role of angiogenesis in the growth and development of tumors. Most cancers depend on the angiogenic process to survive. In general, malignancies are highly vascular in relation to their normal tissue counterparts. Metastatic potential, tumor aggressiveness, overall prognosis, and response to treatment correlate with tumor angiogenesis in a variety of solid tumors. In addition to basic histopathological assessments of angiogenesis, molecular markers of neovascularization are readily apparent in most neoplasms, especially when aggressive phenotypes arise. The process of angiogenesis is the result of a balance between stimulating and inhibiting factors. The most important angiogenic elements, which play various roles in angiogenesis, are VEGF, VEGF receptor 2 (VEGFR-2), and basic fibroblast growth factor (bFGF). Some studies have shown the importance of notch cell-surface receptors and their interaction with various cell membrane ligands (jagged and Dll (Delta-like ligand) molecules) in angiogenesis [2]. Several angiogenic inhibitors serve as important system regulators. These include angiostatin, endostatin, thrombospondin-1, and tumbstatin [2]. Bone marrow–derived stem cells are also important to the development of new vessels [2].

Thus, exploring angiogenic relationships in the pituitary by looking at normal tissue, adenomatous tissue, and adenomatous tissue that has become aggressive may help determine what might make pituitary adenomas unique.

Because angiogenesis plays an important role in the pathogenesis of tumors, the control of angiogenesis represents a rational approach to treatment [3]. In this chapter, we will review the role of angiogenesis in pituitary adenomas.

Angiogenesis and Pituitary Tumors

Abundant angiogenesis in either benign or aggressive pituitary adenomas appears to be generally nonexistent in many of these tumors. In certain subtypes, however, angiogenesis seems to play more of a role in tumor behavior. Discovery of the nonangiogenic mechanisms that cause many of these tumors to be aggressive could shed

Molecular Pathology of Pituitary Adenomas. DOI: 10.1016/B978-0-12-415830-6.00004-4 light on cancer biology as a whole. Therefore, these tumors may be unique and of great interest in studying angiogenesis-dependent and angiogenesis-independent tumors [1].

Histopathology

Many studies have attempted to correlate various clinical, pathological, immunohistochemical, and imaging features to the quantity and/or quality of microvascular density (MVD) as a marker of angiogenesis [4,5].

MVD and angiogenesis markers can then be assessed and compared in normal and neoplastic tissue. In neoplastic samples, vascular analysis may be compared across the various adenomatous subtypes or analyzed according to various tumor factors, such as the level of hormone secreted, presurgical exposure to tumor-related medications, and tumor aggressiveness. Unlike other tumors that seem to show a brisk angiogenic response, pituitary tumors seem to behave differently. Some researchers report that normal pituitary tissue seems to display more vascularity than tissue from adenoma samples [6,7].

In these neoplasms, MVD and angiogenic markers are at low levels. This can be explained by the presence of a dual blood supply: a portal source (hypothalamic-pituitary) and an arterial source [8]. The source of vessels to the neoplastic tissue is not clear, but supply from the arterial source has been reported.

Although these tumors appear to be quantitatively less vascular than expected, it remains conceivable that new vessels could develop within the tumor. These tumors may thrive in conditions independent of angiogenic constraints; in this way, they differ from other tumors. In this setting, they may even provide useful information regarding endogenous inhibitors of angiogenesis and their role in determining the angiogenic phenotype and resulting tumor behavior [1].

Also, unlike many other solid tumors, in which angiogenesis seems to drive the malignancy, tumor aggressiveness in pituitary adenomas often fails to correlate with vascularity or markers of angiogenesis [4,7,9,10]. Some have found results that revealed specific tumor subtypes when they have been analyzed. Invasive macroprolactins (but not GH-secreting tumors) have been found to be more vascular than their noninvasive counterparts. PRL- and ACTH-secreting tumors that could be cured surgically also had a lower MVD than tumors that could not be cured surgically. This was likely related to invasiveness in these tumors [8].

Molecular Biology—Markers of Tumor Angiogenesis

In addition to MVD quantification, much work has gone into studying molecular markers of angiogenesis in pituitary tumors. Similar to MVD results, the majority of these studies have not found a robust angiogenic response in these tumors. Just as MVD likely varies across adenoma subtypes, angiogenic marker expression varies as well. For example, nonfunctioning, ACTH-secreting, and GH-secreting adenomas may display higher levels of angiogenic markers (VEGF, VEGFR-2, and FGF-2) than normal tissue or other tumor subtypes [11].

The utility of markers of angiogenesis as prognostic indicators of pituitary tumors is addressed by several studies investigating MVD and expression of VEGF, endothelial growth factor, FGF, cyclooxygenase-2 (COX-2), and hypoxia-inducible factor-1α (HIF-1α).

The VEGF and VEGFR Family in Tumor Angiogenesis

Much attention has been focused on the VEGF family of growth factors and the receptor tyrosine kinases that mediate their proangiogenic effects. The major mediator of tumor angiogenesis is VEGF-A. VEGF signals mainly through VEGFR-2, which is expressed at elevated levels by endothelial cells engaged in angiogenesis. VEGFR-1 is also involved, but its role has not been elucidated yet.

Findings of VEGF correlations with tumor invasiveness and proliferation are inconsistent, indicating that VEGF may not contribute directly to tumoral invasion, but it may regulate pathways that do increase tumor volume or mediate invasiveness. This notion is supported by the observation that VEGF expression is not strictly associated with endothelium and vessels but is also expressed by adenoma cells as well [12].

This is a likely consequence of the numerous and diverse genetic and epigenetic ways in which VEGF can be induced. Hypoxia, a characteristic of solid tumors, is an important inducer of VEGF. Its effect is mediated through HIF-1α and HIF-2α. It is commonly held that VEGF action is attributable to a paracrine mechanism by tumor cells. These findings explain, at least in part, why elevated VEGF levels in serum, or even tumor tissue, do not predict a benefit in patients receiving drugs that target the VEGF–VEGFR-2 pathway. Tumor cells of many types, including those of hematologic tumors, express VEGFR (especially VEGFR-1) and also produce VEGF, which indicates that VEGF may sometimes act as a direct (cell-autonomous) autocrine growth factor for tumor cells. Furthermore, in some cases, the VEGFRs may be expressed not on the surface of the tumor cell but rather within the cell, where they promote cell survival by an "intracrine" mechanism [13,14].

There are two major ligands for tie-2, angiopoietin-1 (ang-1) and angiopoietin-2 (ang-2). Ang-1 acts as an agonist, whereas ang-2 acts as an antagonist but it can promote angiogenesis, especially in cooperation with VEGF. These angiopoietins act in concert with VEGF to stabilize and mature new capillaries. Blockade of this tie-2/TEK (tyrosine kinase endothelial) pathway has been more difficult than blockade of the VEGF pathway, in part because of the complexity of agonistic and antagonistic ligands for the same receptor and the problems in finding effective and specific drugs against tie-2 or the angiopoietins. However, antibodies and peptide-like antibodies against ang-2 have recently been developed; they can block tumor angiogenesis and tumor growth in preclinical models [14].

Because the vasculature of pituitary tumors seems to differ from other tumors, inhibition of angiogenesis may play a role. Endostatin, a potent angiogenesis inhibitor, has been found to be coelevated with VEGF in most patients with pituitary adenomas [15]. This elevation may account for the findings of diminished angiogenic response in these tumors. VEGF expression differed in the subtypes, thus implicating different mechanisms of VEGF expression and/or action.

The utility of serum VEGF as a marker of pituitary tumor behavior remains unclear. The potential for VEGF as a marker of aggressive tumor behavior in pituitary neoplasms may be small compared with other types of neoplasms because the former often behave as benign, nonmalignant entities.

Fibroblast and Endothelial Growth Factors and Their Receptors

Studies have shown associations between pituitary tumor behavior and the expression of both bFGF, a well-characterized angiogenic growth factor, and its receptor [16–18]. Most investigations of epidermal growth factor receptor (EGFR) expression in various pituitary tumor subtypes have shown it to be higher in functioning compared to nonfunctioning adenomas. EGFR expression was also shown to be significantly higher in ACTH- than in GH- or PRL-producing adenomas, suggesting that EGFR may be involved in the pathogenesis of ACTH adenomas.

Most studies have shown a good correlation between bFGF expression and clinicopathological parameters, including pituitary tumor diameter, invasiveness, and patient outcome. In addition, expression of EGFR in pituitary tumors has often been shown to be a good predictor of tumor invasiveness [12].

Hypoxia-Inducible Factor-1α

HIF-1α is upregulated under hypoxic conditions, and in turn, it upregulates VEGF. This pathway is thought to be involved in the vascularization of tumors growing under hypoxic conditions. Kim et al. [19] found no significant correlation between the expression of VEGF and HIF-1α; their colocalization was seen in only a few cells. Thus, hypoxia-induced VEGF expression may not be an important vasculogenic pathway in pituitary adenomas. Similarly, Vidal et al. [20] showed that HIF-1α expression did not correlate with MVD, thus suggesting that despite HIF-1α–mediated regulation of VEGF in other tumor types, its expression in pituitary tumors may be affected by alternate pathways.

Examination of a series of pituitary tumors for HIF-1α expression showed it to be limited to the nuclei of tumor and endothelial cells. No significant correlation was found between its expression and patient age, sex, or tumor size. With respect to tumor subtype, studies of HIF-1α expression have demonstrated significantly higher levels in GH and PRL adenomas and carcinomas, whereas the lowest levels were detected in ACTH adenomas. The findings of elevated HIF-1α expression in pituitary carcinomas and its decreased expression in ACTH adenomas are of particular interest and highlight the need for further studies into its value as a predictive marker [20].

Cyclooxygenase-2

COX-2, a key enzyme mediating prostaglandin synthesis, is involved in tumor invasiveness and angiogenesis. Its expression in pituitary tumors was demonstrated recently. Increased COX-2 expression was particularly evident in pituitary carcinomas compared with adenomas and normal pituitary, thus suggesting a significant role in tumor progression. Onguru et al. [21] found increased COX-2 expression in functioning tumors compared to nonfunctioning tumors, and both had lower levels of COX-2 than did carcinomas. Bloomer et al. [22] found COX-2 expression in 83% of 30 pituitary tumors. Its expression was significantly associated with that of LH and TSH. In contrast, in a larger series of 164 pituitary tumors, Vidal et al. [23] found GH, PRL, TSH, and female gonadotrophs to express lower COX-2 levels than male gonadotrophs and oncocytic and nononcocytic null cell adenomas. The results of this study should focus analyses on several clinical variables, including the sex of patients with gonadotrophic tumors because gonadotrophs express higher levels of COX-2 than other pituitary neoplasms. However, COX-2 expression did show a strong correlation with MVD. The role of COX-2 in pituitary malignant transformation requires further research because COX-2 is significantly higher in pituitary carcinomas than in adenomas.

Matrix Metalloproteinases

Matrix metalloproteinases (MMPs) are proteolytic enzymes that break down basement membranes and connective tissue, thus facilitating invasive growth. They do so by breaking down an extracellular matrix and selectively remodeling it. In a recent microarray analysis and gene clustering study, Hussaini et al. [24] found a robust, eightfold increase in MMP-9 expression in invasive compared with noninvasive adenomas, a result in keeping with the findings of earlier studies. Several studies established that increased expression and activity of MMP-9 and/or MMP-2 correspond to invasive tumor phenotypes and higher radiological tumor grades. Yet other studies investigating a possible correlation between pituitary tumor invasiveness and MMP-9 expression failed to show an association. Other members of the MMP family, such as MMP-1, MMP-2, and MMP-3, also have been shown to be expressed differentially in pituitary adenomas. The role of MMP as a clinicopathological marker of pituitary adenomas is not established, despite considerable support for this notion. Discrepancies may be rooted in variability in definition of tumor parameters, especially of invasiveness, because it is often defined variably based on radiological, operative, and/or microscopic findings. Nonetheless, MMPs offer much promise as a predictor of tumor behavior. Standardization of approaches to the measurement of MMP levels and activity may clarify some of these contradictory findings [12].

Pituitary Tumor Transforming Gene

The PTTG is expressed in high levels in pituitary adenomas compared to normal pituitary glands [10]. This may be most evident in secretory (versus nonfunctioning) tumors. Among a variety of purported roles, PTTG overexpression may induce transformation of normal cells to neoplastic cells through angiogenic means, such as bFGF expression. In fact, bFGF and VEGF overexpression has been seen in pituitary adenomas, which may depend on PTTG expression.

Several studies also investigated correlations between the expression of PTTG and other markers relevant to pituitary tumors, particularly VEGF and bFGF, because PTTG promotes angiogenesis in many settings [11,18]. A study of 103 pituitary adenomas showed a significant positive correlation between PTTG and VEGF mRNA levels, as well as between PTTG and VEGFR VEGFR2/KDR (kinase domain receptor) expression levels [9]. Regarding PTTG role, it appears that it contributes to the upregulation of VEGF and FGF expression, all of them being overexpressed in various malignancies.

Angiogenesis and Cancer Stem Cells

Studies have identified in tumors a minor population of cells with the characteristics of "tumor-initiating" cancer stem cells. These cells are thought to drive tumor growth and to constitute the seeds of resistance to treatment. It has been suggested that conventional chemotherapy and other types of drugs attack the latter cells, but not the cancer stem cells. The potent tumorigenic properties of cancer stem cells suggest that they may be strongly proangiogenic, and there is some evidence of this feature, which may help to explain some of the effects of tumor-inhibiting anti-angiogenic drugs.

In addition, putative cancer stem cells in brain tumors reside in close proximity to blood vessels in a "vascular niche." Treatment of orthotopic-transplanted gliomas in mice with antibodies to VEGF disrupts the vascular niche and targets the stem cell population. This population expresses high levels of VEGF and thus would be expected to be sensitive to anti-VEGF treatment. Low-dose metronomic chemotherapy may target the population of cancer stem cells or cells that are like cancer stem cells, especially when it is combined with an anti-angiogenic drug such as anti-VEGFR-2 antibodies [2].

The increased use of anti-angiogenic drugs for the treatment of cancer has emerged from decades of extensive basic and clinical research. The clinical benefits of such drugs, however, are relatively modest. Improvements are likely to come from a more thorough understanding of the molecular and cellular mechanisms governing tumor angiogenesis and the response to anti-angiogenic therapies. A number of recent advances promise to bring about such improvements, including new findings in the VEGF and the VEGFR family, discovery of the notch-Dll4 (delta-like ligand 4) signaling pathway in tumor angiogenesis, identification of the mechanisms of resistance to anti-angiogenic drugs, and observations that suggest a role of angiogenesis in the survival and growth of cancer stem cells. These discoveries and others suggest strategies for improving the clinical benefits of anti-angiogenic therapy. These strategies include the development of better preclinical models to study the biology of tumor angiogenesis and anti-angiogenic therapies. Such improvements will also be critical in the use of long-term anti-angiogenic therapy in the adjuvant setting in patients with early-stage disease. With respect to the treatment of metastatic disease, the magnitude and diversity of targets for anti-angiogenic approaches suggest numerous possibilities for anti-angiogenic drug combinations that should be much more effective than monotherapy in treating cancer.

References

1. Cohen, A.B. and Lessell, S. (2009) Angiogenesis and pituitary tumors. *Semin. Ophthalmol.*, 24(3): 185–89.
2. Kerbel, R.S. (2008) Tumor angiogenesis. *N. Engl. J. Med.*, 358(19): 2039–49.
3. Ciccarelli, A., Daly, A.F. and Beckers, A. (2005) The epidemiology of prolactinomas. *Pituitary*, 8(1): 3–6.
4. Niveiro, M., Aranda, F.I., Peiró, G., Alenda, C. and Picó, A. (2005) Immunohistochemical analysis of tumor angiogenic factors in human pituitary adenomas. *Hum. Pathol.*, 36(10): 1090–95.
5. Turner, H.E., Nagy, Z., Gatter, K.C., Esiri, M.M., Harris, A.L. and Wass, J.A. (2000) Angiogenesis in pituitary adenomas—relationship to endocrine function, treatment and outcome. *J. Endocrinol.*, 165(2): 475–81.
6. Di Ieva, A., Grizzi, F., Ceva-Grimaldi, G., Russo, C., Gaetani, P., Aimar, E., et al. (2007) Fractal dimension as a quantitator of the microvasculature of normal and adenomatous pituitary tissue. *J. Anat.*, 211(5): 673–80.
7. Takada, K., Yamada, S. and Teramoto, A. (2004) Correlation between tumor vascularity and clinical findings in patients with pituitary adenomas. *Endocr. Pathol.*, 15(2): 131–39.
8. de la Torre, N.G., Wass, J.A. and Turner, H.E. (2004) Morphologic changes and molecular regulation of angiogenesis in pituitary adenomas. *Front. Horm. Res.*, 32: 133–45.
9. Fukui, S., Nawashiro, H. and Otani, N. (2003) Vascular endothelial growth factor expression in pituitary adenomas. *Acta Neurochir. Suppl.*, 86: 519–21.
10. Fukui, S., Otani, N. and Nawashiro, H. (2003) The association of the expression of vascular endothelial growth factor with the cystic component and haemorrhage in pituitary adenoma. *J. Clin. Neurosci.*, 10(3): 320–24.
11. McCabe, C.J., Boelaert, K. and Tannahill, L.A. (2002) Vascular endothelial growth factor, its receptor KDR/Flk-1, and pituitary tumor transforming gene in pituitary tumors. *J. Clin. Endocrinol. Metab.*, 87(9): 4238–44.
12. Salehi, F., Agur, A., Scheithauer, B.W., Kovacs, K., Lloyd, R.V. and Cusimano, M. (2010) Biomarkers of pituitary neoplasms: A review (part II). *Neurosurgery*, 67(6): 1790–98.

13. Hanahan, D. (1997) Signaling vascular morphogenesis and maintenance. *Science*, 277: 48–50.
14. Oliner, J., Min, H., Leal, J., Yu, D., Rao, S., You, E., et al. (2004) Suppression of angiogenesis and tumor growth by selective inhibition of angiopoietin-2. *Cancer Cell*, 6: 507–16.
15. Gruszka, A., Kunert-Radek, J. and Pawlikowski, M. (2005) Serum endostatin levels are elevated and correlate with serum vascular endothelial growth factor levels in patients with pituitary adenomas. *Pituitary*, 8(2): 163–68.
16. Fukui, S., Otani, N., Nawashiro, H., Yano, A., Nomura, N., Miyazawa, T., et al. (2002) Subcellular localization of basic fibroblast growth factor and fibroblast growth factor receptor 1 in pituitary adenomas. *Brain Tumor Pathol.*, 19(1): 23–9.
17. Qian, Z.R., Sano, T., Asa, S.L., Yamada, S., Horiguchi, H., Tashiro, T., et al. (2004) Cytoplasmic expression of fibroblast growth factor receptor-4 in human pituitary adenomas: Relation to tumor type, size, proliferation, and invasiveness. *J. Clin. Endocrinol. Metab.*, 89(4): 1904–11.
18. McCabe, C.J., Khaira, J.S., Boelaert, K., Heaney, A.P., Tannahill, L.A., Hussain, S., et al. (2003) Expression of pituitary tumour transforming gene (PTTG) and fibroblast growth factor-2 (FGF-2) in human pituitary adenomas: Relationships to clinical tumour behaviour. *Clin. Endocrinol. (Oxf.)*, 58(2): 141–50.
19. Kim, K., Yoshida, D. and Teramoto, A. (2005) Expression of hypoxia-inducible factor 1alpha and vascular endothelial growth factor in pituitary adenomas. *Endocr. Pathol.*, 16(2): 115–21.
20. Vidal, S., Horvath, E., Kovacs, K., Kuroki, T., Lloyd, R.V. and Scheithauer, B.W. (2003) Expression of hypoxia-inducible factor-1alpha (HIF-1alpha) in pituitary tumours. *Histol. Histopathol.*, 18(3): 679–86.
21. Onguru, O., Scheithauer, B.W., Kovacs, K., Vidal, S., Jin, L., Zhang, S., et al. (2004) Analysis of Cox-2 and thromboxane synthase expression in pituitary adenomas and carcinomas. *Endocr. Pathol.*, 15(1): 17–27.
22. Bloomer, C.W., Kenyon, L., Hammond, E., Hyslop, T., Andrews, D.W., Curran, W.J., et al. (2003) Cyclooxygenase-2 (COX-2) and epidermal growth factor receptor (EGFR) expression in human pituitary macroadenomas. *Am. J. Clin. Oncol.*, 26(4): S75–80.
23. Vidal, S., Kovacs, K., Bell, D., Horvath, E., Scheithauer, B.W. and Lloyd, R.V. (2003) Cyclooxygenase-2 expression in human pituitary tumors. *Cancer*, 97(11): 2814–21.
24. Hussaini, I.M., Trotter, C., Zhao, Y., Abdel-Fattah, R., Amos, S., Xiao, A., et al. (2007) Matrix metalloproteinase-9 is differentially expressed in nonfunctioning invasive and noninvasive pituitary adenomas and increases invasion in human pituitary adenoma cell line. *Am. J. Pathol.*, 170(1): 356–65.

5 Apoptosis

Pituitary adenomas are known to vary from the standpoint of their intrinsic aggressiveness, which might imply that a different therapeutical approach could be useful. The morphological markers positively correlated with aggressiveness in most tumor systems are of limited usefulness in pituitary neoplasms [1]. The biological behavior of pituitary tumors depends on a number of factors, including growth rate, which in turn depends on the number of cycling cells versus cells dying via necrosis or apoptosis [2]. Cell proliferation activity seems to be prognostically useful in anterior pituitary adenomas.

The term "apoptosis" comes from a Greek word, and it refers to the falling off of the petals of the flowers and hyperplastic and neoplastic lesions. Also called "programmed cell death," the mechanism can be described as consisting in several active processes controlling tissue homeostasis and carcinogenesis; the balance between proliferating and apoptotic cells plays an important role in the growth rate of pituitary tumors and might be a target for antitumor therapy [3–8].

Detection of Apoptosis

Pituitary adenoma cells undergoing apoptosis exhibit a common prototypical pathway of changes. Apoptosis can also be detected through DNA and biochemical assays based on the specific pattern of nucleosomal fragmentation.

Intranucleosomal DNA cleavage, in approximately 180-bp segments, offers an important target for apoptosis detection; the derived DNA strand breaks have many new 3′-OH ends. The TUNEL technique is based on the incorporation of labeled deoxyribonucleotide triphosphates by terminal deoxynucleotidyl transferase (TdT). The *in situ* end-labeling (ISEL) technique, which uses tailing of the 3′-OH ends by the Klenow DNA polymerase, represents an alternative assay [3].

Although the results are not uniform, it has been proved that apoptosis occurs with low frequency in a subset of pituitary adenomas. However, apoptosis in its early stages can be identified using alternative techniques based on remodeling the cytoskeleton by caspase activity [3].

Apoptosis Versus Necrosis

Apoptosis is a process that eliminates damaged single cells rapidly without causing any disturbance in the environment. It is distinctly different from ordinary necrosis, which is a passive and non-ATP-dependent lethal process caused by a general failure of cellular homeostasis due to external factors which typically involves large groups

Molecular Pathology of Pituitary Adenomas. DOI: 10.1016/B978-0-12-415830-6.00005-6

of cells. In necrosis, the cell swells and bursts, spilling out its contents and often provoking an inflammatory response [3].

Regulation and Execution of Apoptosis

Generally, most cells require growth factors and mitogen signals to survive, and deprivation of these factors may lead to apoptosis [3]. Apoptosis induction can be activated by two pathways. In the extrinsic pathway, apoptosis is triggered by extracellular ligands acting on transmembrane (TM) "death receptors," such as Fas ligand (FasL) and TNF (tumor necrosis factor) [9]. FasL binding induces clustering of Fas "death domains"; subsequently, a cytoplasmic protein called Fas-associated death domain, which also contains a death domain, binds to the intracellular portion of Fas. The complex recruits procaspase-8 molecules, activated caspase is generated by autoproteolysis, leading to downstream activation of other caspases (such as caspase-3, caspase-6, and caspase-7) and cell destruction.

Fas-dependent apoptosis can be modified by Fas receptor glycosylation or by transcriptional upregulation of Fas expression by the activated p53 protein. The intrinsic pathway of apoptosis (mitochondrial) operates in response to various types of intracellular stress: growth factor withdrawal, DNA damage, unfolding stresses in the endoplasmic reticulum, and death receptor stimulation. Following the reception of stress signals, pro-apoptotic bcl-2 family proteins are activated and subsequently inactivate anti-apoptotic bcl-2 proteins. This interaction destabilizes the mitochondrial membrane and releases apoptotic factors that trigger the caspase proteolytic cascade, chromatin condensation, and DNA fragmentation. The key players in the intrinsic pathway are the bcl-2 family of proteins, which are critical death regulators residing immediately upstream of mitochondria.

The bcl-2 gene is a member of a large family that includes genes that can inhibit or promote apoptosis [3]. The bcl-2-related proteins share homology in four highly conserved domains referred to as BH1–BH4. These domains serve to homodimerize and heterodimerize and, thus, influence cell death or cell cycle entry. Most bcl-2 family members possess another domain called the TM domain; deletion of the TM domain renders both bcl-2 and bax incapable of completing their anti- or pro-apoptotic function.

The mitochondrion, which is the focus of the actions of the bcl-2 family, is intimately involved in the delicate network of apoptosis pathways capable of regulating apoptotic signals. The apoptotic protease-activating factor-1 (APAF-1) is capable of binding to procaspase-9, subsequent to cytochrome *c* binding. Binding of APAF-1 with cytochrome *c* and procaspase-9 forms the holoenzyme apoptosome. The apoptosome is capable of activating caspase-3 and other effector caspases that are required for the final stages of apoptotic cell death [3,10].

The bcl-2 family of proteins regulates various steps in apoptosis. Some of the members of this gene family, including bcl-2, bcl-XL, and Mcl-1, block cell death, whereas others, such as bax, bad, and bcl-Xs, promote programmed cell death.

Overexpression of bcl-2 provides protection against various apoptotic stimuli, including growth factor deprivation, radiation, chemotherapeutic drugs, oncogenes such as c-myc, and tumor suppressor genes such as p53 [7]. Sambaziotis et al. [10] have found that bcl-2 and bax molecules play a role in the regulation of apoptotic mechanisms in pituitary adenomas [11].

The bcl-2 gene was found to contribute to the neoplastic progression by inhibiting apoptosis induced by hormones and cytokines, thus extending cell survival [10].

The bax gene can dimerize within itself or with bcl-2, and when overexpressed, bax homodimers promote apoptosis. When bcl-2 is in excess, bcl-2 homodimers predominate and cells are protected from death. The bcl-2 gene inhibits apoptosis without increasing cell proliferation, and it negatively regulates the apoptotic activity of bax by the formation of bcl-2/bax heterodimers. The bcl-2/bax ratio represents a cell-death switch, which predetermines cell life or death response to an apoptotic stimulus [10].

Factors causing DNA damage are able to induce apoptosis via activation of p53. Depending on the severity of DNA damage and the cell type, p53 will either cause cell cycle arrest or activate the apoptotic cell destruction [3]. The pro-apoptotic gene bax has been shown to be a p53 target and is upregulated in certain cell types during p53-mediated apoptosis.

The members of the caspase family play critical roles in the regulation and execution of apoptosis. They can be divided into two groups according to their function: initiator (or upstream) caspases, which are activated by pro-apoptotic signals, and effector (or downstream) caspases, which are activated by the initiator ones. Effector caspases carry out the proteolytic cell destruction and produce most of the morphological changes of apoptosis. Caspases normally exist in the form of proenzymes (procaspases) and are converted to the entirely active form by sequential proteolytic cleavage, which is usually performed by caspases themselves. Once a caspase is activated, the effect can be multiplied exponentially by sequential activation of other procaspases. This pattern is called the caspase cascade, and it is widely used for activating the effector caspases caspase-3, caspase-6, and caspase-7 [12]. One of these (caspase-3) is responsible for the activation of the DNase (deoxyribonuclease), which performs the fragmentation of DNA at the end stage of apoptosis [3].

Apoptosis in Human Pituitary Adenomas

Because of the limited number of studies examining pituitary adenomas, very little is known about the regulation and significance of apoptosis [7,13–16]. Pituitary adenomas are neoplasms with a low proliferation rate, as evidenced by the absence of mitotic activity and infrequent mitoses even in aggressive tumors, and by a low apoptotic index, S-phase fraction, and Ki-67 proliferation index. Apoptosis and mitoses are asynchronous, opposed events representing two sides of the same coin. Conceptually, in neoplastic conditions, the balance between mitoses and apoptosis is altered in favor of mitoses. Even though both events are commonly upregulated, only scarce mitoses can be identified in pituitary adenomas. Therefore, the number of apoptotic cells is also expected to be very small. The increased mitotic index in pituitary carcinomas compared to adenomas was found to correlate with an increased apoptotic index [7]. Therefore, due to very low apoptotic activity in pituitary adenomas, it is difficult to detect apoptosis based on histology alone. This task is even more difficult in electron microscopic studies due to the low number of cells. EM represents the best tool to illustrate apoptosis with accuracy, particularly to study in detail the sequence of the apoptotic process.

Morphology of Apoptosis in the Human Pituitary

In routine histological stains, cells display shrinkage, with marked reduction of volume. The early apoptotic nuclei show gradual pyknosis due to chromatin condensation, often with a slightly irregular contour. Subsequently, they show margination of chromatin with typical crescent formation. The cytoplasm appears compact, the cells gradually lose connections with adjacent cells, and finally, they appear to be floating. Nuclear fragmentation into small, highly pyknotic particles with formation of apoptotic bodies represents advanced events of this process.

By EM, pituitary adenoma cells undergoing apoptosis exhibit a common prototypical morphologic pathway of changes that can be divided into three phases [3]. The first phase is characterized by prominent nuclear alterations. The sequence of these changes includes discrete clumping and condensation of chromatin with margination along the nuclear membrane and crescent formation or accumulation within one pole of the nucleus. Subsequently, the nucleus is separated from the cytoplasm by a perinuclear halo. In the early apoptotic stage, cytoplasmic organelles such as mitochondria and Golgi apparatus remain well preserved, whereas the endoplasmic reticulum may be dilated. The morphology and distribution of secretory granules appear normal. Reorganization of the subplasmalemmal area and the increase in electron opacity of the cytoplasmic matrix represent the most striking cytoplasmic change observed during the early stage of apoptosis.

In the second phase, both nuclear and cytoplasmic changes are striking. The condensed nucleus breaks into smaller, dense fragments, which contain highly electron-dense chromatin mass. Typically, the cells become round and the cytoplasm is extensively altered, with marked accumulation of vacuoles of varying sizes. A few intact mitochondria and secretory granules are present. Subsequently, at the "blebbing" stage, the apoptotic cells are detached from the adjacent cells. Their outlines become convoluted with extensions, and the plasma membrane surrounds the detached, solid cellular material, forming apoptotic bodies. The latter are crowded, closely packed, and highly condensed nuclear fragments, which lose their membranes and are encircled by a narrow cytoplasmic rim.

In the third phase, the apoptotic bodies are eliminated rapidly via phagocytosis, mostly by macrophages and stellate cells. Remnants of degraded apoptotic bodies can be identified within these cells. Infrequently, apoptotic bodies are not eliminated by phagocytosis, but they undergo degradation in a sequence known as "secondary necrosis" [16]. Among the different types of pituitary adenomas, corticotroph adenomas show the most frequent and striking apoptotic changes at the electron microscopic level [15].

A rare cell-death process, distinct from apoptosis and commonly occurring in dopamine-treated, PRL-producing adenomas and in oncocytomas, can also be observed. The main feature of this process is the progressively increased density of the cytoplasmic matrix, resulting in the overall appearance of a "dark" cell. The nuclei show no fragmentation, the chromatin becomes highly electron-dense, retaining its original distribution even at advanced stages of development. Subsequently, the affected cells undergo severe alterations in cytoplasmic organelles and show extensive vacuolization while maintaining the integrity of the nucleus. Dark cells can be observed concurrently with apoptotic cells in the same tissue section.

Initial immunocytochemical studies have found that approximately one-third of pituitary adenomas show abnormal expression of bcl-2, and that most of these tumors coexpressed c-myc oncoproteins as well [3]. In addition, bcl-2 was found to occur in all adenoma types, but less frequently than bax. Nonfunctioning tumors were more often positive for bcl-2. All functioning adenomas and the substantial majority of nonfunctioning tumors were immunopositive for bax, and bax was less frequently positive in nonfunctioning tumors than bcl-2 [3].

The wider expression of bax than bcl-2 supports the suggestion that this molecule has additional functions besides inhibiting bcl-2 [7]. The few reported studies of apoptosis in human pituitaries have simply described the detection of apoptotic cells in various types of pituitary adenomas or in primary cultures of human pituitary adenoma cells.

Apoptosis is observed frequently in hormone-dependent cells, usually after hormonal withdrawal.

In a study of 85 adenomas, Kontogeorgos et al. [6] found 54% apoptotic labeled nuclei. The apoptotic LI (ALI) has a significantly higher function than nonfunctioning adenomas.

ALI was also investigated using the TUNEL technique by Losa et al. [13], and no significant differences between microadenomas and macroadenomas were observed. In contrast, study by Kontogeorgos et al. [6] regarding functional group of tumors revealed a higher ALI in microadenomas than in macroadenomas without statistically significant differences; however, untreated microadenomas, particularly lactotroph adenomas, showed a higher ALI.

However, Kulig et al. [7] reported that preoperative treatment of adenomas with octreotide or dopamine agonists did not change the apoptotic index significantly. They also found an ALI that was four times higher in pituitary carcinoma than in adenomas. Pituitaries from pregnant and postpartum women had an ALI that was five times higher than matched controls from nonpregnant females.

To investigate the effects of octreotide administration on the growth rate of GH-secreting pituitary adenomas, Losa et al. [14] measured both the Ki-67 LI and the ALI in tumor specimens from octreotide-treated or matched untreated acromegalic patients. Overall, the mean Ki-67 LI of treated patients was 53% lower than that in untreated patients. The antiproliferative effect of octreotide occurred independent of tumor extension and invasiveness. Octreotide-treated and untreated patients showed similar ALIs. There was a positive correlation between the Ki-67 LI and the ALI. Their study demonstrates that acromegalic patients receiving chronic octreotide treatment have a lower value for the proliferation marker Ki-67, but no significant difference in the ALI compared with matched untreated patients. The antiproliferative effect of octreotide on GH-secreting adenomas should imply a lower risk of tumor growth during long-term chronic treatment with the drug.

Somatostatin plays an important role in the regulation of GH secretion in humans. The effect of somatostatin is mediated through high-affinity, G protein–coupled membrane receptors that are expressed in all somatostatin-targeted organ systems.

The efficacy of octreotide treatment is still to be established, considering its potential antiproliferative effect on GH-secreting adenomas. The rate of growth of pituitary tumors depends on the balance between the proliferating cells and the loss of tumor cells by apoptosis (programmed cell death) and ischemic or hemorrhagic events. Few studies investigated this issue in GH-secreting adenomas; octreotide-treated acromegalic patients showed a significant reduction of the Ki-67 LI compared with untreated patients, whereas no difference in the ALI was detected in similarly treated patients. However, no study reported the results of both proliferation and ALIs in the same tumors [14].

Our results show a positive correlation between proliferative activity and tumoral invasiveness; the mean value of the Ki-67 LI was significantly higher in invasive adenomas relative to noninvasive adenomas. Some differences were also recorded in functioning and nonfunctioning adenomas. By the ISEL technique, apoptosis was present in 60% of the tumors. By flow cytometry analysis, apoptosis was detected in functioning and nonfunctioning tumors with no significant differences. This phrase should be completely omitted, not being relevant. The ALI was somewhat lower than the TUNEL-detected index.

In the majority of the studied invasive tumors, we found a high number of cells in the S/G2/M phases, which can be associated with the aggressive behavior of the tumors. The assessment of apoptosis and proliferation might contribute to a better understanding of the regulatory mechanisms implicated in tumor progression, and thus provide guidance to a specific therapeutic approach. Flow cytometry is a valuable clinical laboratory tool to perform a fast and valid analysis of apoptosis and cell cycles in pituitary tumors [17] (Figure 5.1).

We have also found that APAF-1 seems to be downregulated in most of the invasive pituitary tumors. In the other cases (low-grade invasive tumors), it was expressed in variable degrees, with a zonal distribution. By contrast, in invasive pituitary tumors, cathepsin-B was commonly expressed with a granular pattern, mostly lying on the plasma membrane. We noticed a negative correlation between APAF-1 and invasiveness and a positive correlation between cathepsin-B and invasiveness. There is also an inverse correlation between APAF-1 and cathepsin-B expressions. A bidirectionally inverted relation between APAF-1 and cathepsin-B expressions has been observed. A shifting balance between cell-death mediators might result in changes in tumor behavior [8].

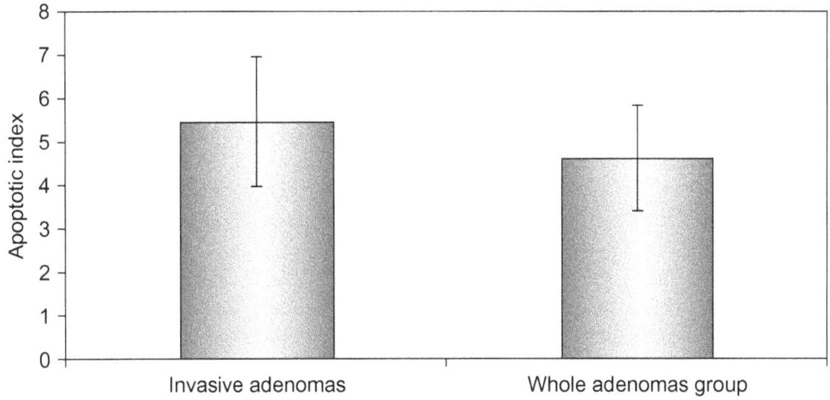

Figure 5.1 Apoptotic index in invasive adenomas and the whole adenomas group.

A combination of various methods provides a more reliable method for the evaluation of apoptosis in pituitary adenomas. A better insight can be given by the genomic and proteomic approaches, in order to elucidate the complicated pathways involved in the regulation of apoptosis in pituitary adenomas.

References

1. Buchfelder, M., Fahlbusch, R., Adamse, F., Kiesewetter, F. and Thierauf, P. (1996) Proliferation parameters for pituitary adenomas. *Acta Neurochir.*, 65: 18–21.
2. Abdel-Moneim, I., Melamed, M.R., Darzynkiewicz, Z. and Gorczyca, W. (2000) Proliferation and apoptosis in solid tumors. *Anal. Quant. Cytol. Histol.*, 22(5): 393–97.
3 Kontogeorgos, G., and Kovacs, K (eds) (2004) *Molecular Pathology of the Pituitary.* Karger, Basel.
4. Schwartz, L.M. and Osborne, B.A. (1993) Programmed cell death, apoptosis and killer genes. *Immunol. Today*, 14: 582–90.
5. Fornas, O., Mato, M.E., Viader, M., Oliver, B., Tresserres, P., Bartumeus, F., et al. (1996) Flow cytometry analysis of pituitary adenomas. *Horm Res.*, 46(6): 257–62.
6. Kontogeorgos, G., Sambaziotis, D., Piaditis, G. and Karameris, A. (1997) Apoptosis in human pituitary adenomas: A morphologic and *in situ* end-labeling study. *Mod. Pathol.*, 10(9): 921–26.
7. Kulig, E., Jin, L., Qian, X., Horvath, E., Kovacs, K., Stefaneanu, L., et al. (1999) Apoptosis in nontumorous and neoplastic human pituitaries—expression of the Bcl-2 family of proteins. *Am. J. Pathol.*, 154: 767–74.
8. Tanase, C., Codorean, E., Ardeleanu, C., Butur, G., Manda, G., Neagu, M., et al. (2004) Apoptosis in pituitary adenomas; Comparative study by flow cytometry and *in situ* end-labeling technique. In *Second Intercontinental Congress of Pathology, Iguassu Falls, Brazil, June 9–13* (Soares, F., Vassallo, J., and Torres, L.F.B., eds). Medimond International Proceedings, Medimond—Monduzzi Editore, Italy, pp. 117–120, http://www.monduzzi.com.
9. Schmitz, I. and Kirchkoff, P.H. (2000) Regulation of death receptor-mediated apoptosis pathways. *Int. J. Biochem. Cell Biol.*, 32: 1123–36.

10. Sambaziotis, D., Kapranos, N. and Kontogeorgos, G. (2000) Bcl-2 gene family in endocrine pathology: A review. *Endocr. Pathol.*, 11: 205–13.
11. Tamm, I., Schriever, F. and Dorken, B. (2001) Apoptosis: Implications of basic research for clinical oncology. *Lancer Oncol.*, 2: 33–42.
12. Reed, J.C. (2000) Mechanisms of apoptosis. *Am. J. Pathol.*, 157: 1415–30.
13. Losa, M., Barzaghi, R.L., Mortini, P., Franzin, A., Mangili, F., Terreni, M.R., et al. (2000) Determination of the proliferation and apoptotic index in adrenocorticotropin-secreting pituitary tumors: Comparison between micro- and macroadenomas. *Am. J. Pathol.*, 156: 245–51.
14. Losa, M., Ciccarelli, E., Mortini, P., Barzaghi, R., Gaia, D., Faccani, G., et al. (2001) Effects of octreotide treatment on the proliferation and apoptotic index of GH-secreting pituitary adenomas. *J. Clin. Endocrinol. Metab.*, 86: 5194–200.
15. Vidal, S., Horvath, E., Kovacs, K., Scheithauer, B.W., Lloyd, R.V. and Kontogeorgos, G. (2001) Ultrastructural features of apoptosis in human pituitary adenomas. *Ultrastruct. Pathol.*, 25: 85–92.
16. Messini, I., Kontogeorgos, G., Kovacs, K., Horvath, E. and Smyth, H.S. (2002) Pituitary corticotroph adenoma containing many apoptotic cells: A histologic, immunohistochemical, ultrastructural and molecular study. *Ultrastruct. Pathol.*, 26: 27–32.
17. Tanase, C., Codorean, E., Arsene, D., Ardeleanu, C., Raducan, E., Popescu, I.D., et al. (2008) APAF-1 and cathepsin-B, regulators of mitochondrial apoptotic pathway in pituitary tumor progression. *Histopathology*, 53: 219.

6 Signal Transduction

Signaling Molecules and Pathways Involved in Pituitary Tumorigenesis: Diagnostic Tools

In pituitary tumors, the most-studied and investigated cell signaling pathways involved in tumorigenesis are MAPKs, AKT, Ras, and Wnt intracellular cascades. Both AKT and MAPK pathways are enhanced in many pituitary tumors, leading to downregulation of physiological inhibitors of the cell cycle. These pathways interconnect at the level of intracellular signaling starting from TRKs and growth factor receptors and may constitute an early event in tumorigenesis [1]. Disturbances have been reported in pituitary cell transformation at each level that links receptors to intracellular pathways and finally to development of a certain cellular function.

Although they are interconnected, we will present receptors and signaling pathways separately. In pituitary tumors, therapeutic approaches aiming signaling molecules as possible therapeutic targets are also highlighted.

Receptors Involved in Pituitary Transformation

Tyrosine-Kinase Receptors

A large number of ligands—such as FGF2, IGF1, TGFα, EGF, and VEGF—are present in the environment of pituitary cells, either normal or adenomatous, and are acting through their specific receptors, belonging to the class of protein tyrosine kinases. Ligand-induced receptor activation by autophosphorylation occurs rapidly, leading to a short-lived active form (reversed by phosphoprotein phosphatases). A signaling cascade that involves scaffolding proteins and an array of MAPKs generates a set of longer-life transducers and selectively guides the signal toward intracellular targets, including at the nuclear level.

Growth and angiogenic factors are triggering molecules activating various signaling pathways highly involved in pituitary tumorigenesis. FGF2 and FGF4 have been the most-studied growth factors in relation to pituitary tumor promotion [2]. Reports focusing on high levels of FGF2 in the pituitary led to the suggested role in growth and differentiation. In addition, transforming FGF4 sequences have been isolated from human PRL-secreting pituitary tumors, suggesting their involvement in pituitary tumor progression [2].

Molecular Pathology of Pituitary Adenomas. DOI: 10.1016/B978-0-12-415830-6.00006-8

Elevated levels of angiogenic factors, like VEGF, have been reported in patients diagnosed with all types of pituitary tumors. Both FGF2 (bFGF) and VEGF expressions correlate with pituitary tumor size and aggression [2]. Moreover, some angiogenic factors have HIF-responsive elements on their promoters [3].

TGFα is one of several growth factors that bind the EGFR, and is overexpressed in pituitary tumors. In pituitary adenomas, bFGF and TGFα are expressed in all cell types and are downregulated by TGFβ3.

Neighboring cells, such as folliculostellate cells (FSCs), create a microenvironment for pituicytes by secreting an array of growth factors, especially EGF and cytokines. The abnormal growth of pituicytes is highly mediated by paracrine mechanisms in which a central role is the tandem EGFR–PTTG. A future therapy of pituitary adenomas could benefit from the inhibition of EGFR-mediated PTTG1 expression by intracellular blockade of paracrine signaling [4].

As epidermal and fibroblast growth factors are overexpressed in human pituitary tumors, all downstream components of their signaling pathways are also deregulated.

EGFR pathway substrate 8 (Eps8) transcripts are overexpressed in human pituitary tumors compared to normal pituitary. The activation pathway triggers increased proliferation and anti-apoptotic mechanisms in pituitary cells by activation of MAPK pathways via ERK (Extracellular signal-regulated kinases), and of phosphoinositol-3 kinase (PI3K) via AKT. All these signaling molecules were overexpressed in pituitary tumors. Consistently increased levels of Eps8 protein, activated AKT, and activated ERK were found, especially in gonadotrophinomas.

Because Eps8 can amplify growth factor receptor signaling to promote proliferation and cell survival in human pituitary tumors, silencing Eps8 blocks pituitary cell proliferation. Eps8 overexpression may be associated with other tumor-inducing mutations that result in growth factor and/or growth factor receptor upregulation, which is reported to contribute to the invasiveness of pituitary tumors. Upregulation of these signaling molecules supports the functional significance of Eps overexpression in human pituitary tumorigenesis [5].

Considering only growth and angiogenic factors and their receptors, it appears that the key signaling molecules related to pituitary adenomas are bFGF, VEGF, REK (recombinant enterokinase), stromal cell-derived factor (SDF), and Eps8. These molecules can be good candidates for both diagnosis and targeted therapy.

G-Protein–Coupled Receptors

G-protein–coupled receptors (GPCRs) respond to extracellular stimuli, like hormones, by interaction with a G-protein, transducing a signal across the membrane into the cellular interior. After GPCR activation, Gα-subunits bind GTP and become active, further activating downstream signaling factors like the enzyme adenylyl cyclase (AC), which synthesizes cyclic-AMP (cAMP). Activated G-proteins interact with downstream signaling factors to alter the production of second messengers such as inositolphosphates, calcium, and cAMP. GPCRs that activate the Gi class of Gα subunits inhibit cAMP production and GPCRs that activate the Gs class of Gα subunits activate cAMP production. cAMP, in turn, activates the cAMP-dependent protein kinase, protein kinase A (PKA). The PKA activation pathway is an example of a signal transduction cascade, in which tying several signaling events together amplifies the original signal inside the cell. For each activated GPCR molecule, many G-proteins can be activated, and each active G-protein can synthesize many cAMP molecules, continuing the cascade to PKA and further downstream.

GPCRs are highly involved in pituicyte functions. Impaired signaling was demonstrated in pituitary adenomas. Gsα point mutations have been demonstrated in GH-secreting pituitary tumors [6].

Pertuit et al. [7] have widely analyzed Gsα alterations in GH-secreting adenomas.

The only mutations so far unequivocally identified and observed in 30–40% of the GH-secreting tumors concern the gsp oncogene. Despite the large individual variations of Gsα mRNAs [7], the level of Gsα proteins is always lower in gsp positive (gsp+) compared to gsp negative (gsp−) tumors [7,8]. It has been previously suggested that activation of Gsα induces a conformational change that prevents its attachment to membranes and increases its degradation rate, which could involve the proteasome [9].

In addition to the gsp oncogene, an overexpression of the WtGsα protein has been observed in a subset of gsp− adenomas. Approximately 60% of these gsp− tumors express high levels of Gsα compared to normal human pituitary cells. The GNAS locus (guanine nucleotide binding protein (G protein), alpha stimulating activity polypeptide 1), which maps on human chromosome 20q13, consists of a complex region with multiple alternative spliced transcripts encoding multiple protein products. In most human tissues, Gsα is biallelically expressed but, in specific tissues, Gsα is imprinted [10,11]. In pituitary tumours, Gsα-encoding transcripts are monoallelically expressed, predominantly from the maternal allele [12]. In almost all cases of gsp+ somatotroph adenomas, the GNAS-activating mutation occurs on the active maternal allele [12]. It is well known that genomic imprinting dysregulations can impact gene expression levels and so can participate in tumorigenesis. A strong imprinting relaxation, with a paternally derived expression of Gsα, has been found only in gsp− tumors. Thus, other mechanisms that could account for WtGsα overexpression remain to be identified.

Gs proteins couple hormonal stimulation of various cell-surface receptors to the activation of AC [7]. AC activation leads to the generation of intracellular second-messenger cAMP, which stimulates the PKA, the main cAMP effector. The phosphodiesterases (PDEs) contribute to the complexicity and specificity of the cAMP pathway by hydrolyzing cAMP. It is now well established that cAMP is compartmentalized in cells. In response to an elevation of cAMP, PDEs can be activated directly by PKA (i.e., rapid feedback regulation) and/or by induction of PDE gene transcription (i.e., long-term regulation) [13]. Thus, the spatiotemporal balance between PKA and PDE activities is a determinant in the control of the cAMP signaling.

In the absence of PDE inhibitors, no difference in intracellular cAMP levels between gsp+ and gsp− adenomas was detectable [7]. Persani et al. [14] demonstrated that the transcripts of PDE4C and 4D, as well as those of PDE8, were overexpressed in gsp+ tumors [13], which were correlated with a sevenfold increase in

PDE activity. The two nuclear proteins, the CREB protein and the inducible cAMP early repressor (ICER), are the main and best-characterized final targets of cAMP. The mRNA levels of the transcription factors CREB and ICER are both increased in the gsp+ tumors [15]. The phosphorylated CREB levels are similar in the two types of tumors, although PDE blockade induces an increase in P-CREB (PhosphoCREB) in gsp+ tumors. These results suggest that an increase in PDE activity could counteract the activation of the cAMP pathway and may have an impact on the phenotype of the gsp+ tumors [13].

Besides alterations of the cAMP pathway in gsp+ tumors, several lines of evidence also suggest the existence of cAMP pathway alterations in GH-secreting adenomas overexpressing WtGsα. Relatively high levels of CREB or ICER mRNAs have been observed in some gsp− tumors [15].

The overexpression of WtGsα enhances intracellular cAMP accumulation and stimulates the cAMP pathway (P-CREB level). An increase in CREB-dependent transcription is also observed both in the presence of the gsp oncogene and with the overexpression of WtGsα in GH3 cells [7].

In order to accurately determine the role of Gsα alterations in the initiation and progression of the GH-secreting adenomas, Pertuit et al. [7] realized a study on pituitary cells, finding that the induction of the expression of the gsp oncogene initiates a considerable increase in the AC activity, which is associated with an increase in the intracellular cAMP level. A weak but long-lasting activation of the AC, associated with a slight increase in the cAMP level, is also observed in response to overexpression of WtGsα. cAMP progressively decreases despite continuous transgene expression, suggesting a potential involvement of the PDEs. This may represent a second mechanism of feedback in addition to the posttranscriptional regulation of the gsp oncogene [7,16].

These mutations inhibit Gsα GTPase activity, resulting in GHRH ligand-independent constitutive activation of cAMP, which results in GH-transcriptional activation and somatotroph proliferation via a CREB in the GH promoter [2].

Significantly, higher amounts of Ser133phosphorylated, and hence activated CREB, have been reported in some GH-secreting pituitary tumors compared with the levels found in nonfunctional (NF) tumors. This augmented CREB activity was evident even in tumors that did not manifest a Gsα mutation. This would suggest that CREB activation may occur via a Gs-independent mechanism [2]. It is possible that the stimulatory/inhibitory polypeptides and steroid hormones released by hypothalamus and peripheral endocrine organs could alter pituitary gene expression and hormone secretion [2].

GHRH and Somatostatin

Although some pituitary tumors express a truncated GHRH receptor (GHRHR), no abnormalities leading to constitutive GHRHR activation have been identified [2,17]. The first molecular alteration in pituitary tumors was a mutation of a G-protein transducing GHRHR signaling. This gsp mutation impairs glycosylation-mediated receptor processing, maturation, ligand binding, and signaling [18].

Pituitary tumors heterogeneously express five somatostatin receptor subtypes (SSTR1–5) [2,19]. Under normal circumstances, somatostatin limits cell growth by acting on the PI3K/AKT signaling pathway. It was demonstrated that somatostatin analogs, such as octreotide [20] or lanreotide [13], can induce antiproliferative action in pituitary tumors by inducing the expression of the tumor suppressor gene Zac1 [20]. Moreover, it was suggested that the antiproliferative effect of somatostatin can occur by downregulation of pERK and upregulation of p27 leading to cell cycle arrest [21].

Somatostatin and dopamine receptors were identified as future targets in pituitary adenomas. Chimerical compounds that could be used for triggering intracellular pathways toward apoptosis activation or other cytostatic functions were proposed. Therefore, the hybrid somatostatin–dopamine compounds could induce apoptosis in pituitary tumors by activating G-proteins that phosphorylate Jun N-terminal kinase (JNK) and, by parallel circuits, increase p53 and decrease Ki-67 [22].

Gonadotrophin-Releasing Hormone

Pituitary adenomas express both GnRH and GnRH receptors. However, as in the case of the GHRH and TRH receptors, no activating GnRH receptor mutations have been described in pituitary tumors [2]. GnRH induces selective signaling via PKC/MAPK substrates, a process that activates specific genes in various GnRH target cells and elicits specific biological end points [23] (Figure 6.1). The same GnRH type I receptor mediates both physiological pathways in normal and in neoplastically transformed pituitary. GnRH ligands (GnRHI and GnRHII, agonists and antagonists of the GnRH type I receptor) may bind active conformations of the receptor that are stabilized by receptor-mediated intracellular complexes in a "cell context"–dependent manner [23].

MAPKs (ERK, JNK, and p38) mediate the pituitary actions of GnRH [24] but also mediate the apoptotic and antiproliferative actions of GnRH in cancer cells [25–28]. It appears that divergent signaling was found owing to different MAPK substrates [23].

Other GPCRs studied in pituitary cells by Massa et al. [29] identified CXCR4 (chemokine receptor 4) and its ligand, a possible novel growth factor: SDF-1, belonging to the CXC (CXC motif chemokine) subfamily of chemokines. The study demonstrated that this molecule is a powerful proliferative factor in pituitary cells, contributing to the regulation of pituitary function and possibly participating in the genesis of pituitary adenomas.

Upregulation of GHRHR at the transcriptional level, which in turn modulates pituitary tumor growth, can be done by reversion-inducing, cysteine-rich protein with kazal motifs (RECK). This molecule interacts directly with MMP2 and MMP9, as an MMP endogenous inhibitor. These reported results indicate RECK as a promising target for the treatment of pituitary adenomas [3]. MMP2 and MMP9 were reported to be both involved in pituitary tumor invasiveness and correlated with angiogenesis and invasiveness [30–32]. Recently, a novel membrane-localized estrogen receptor (ER) has been identified, namely GPR30 (a GPCR), in several cancer cell lines and animal pituitary glands [33,34].

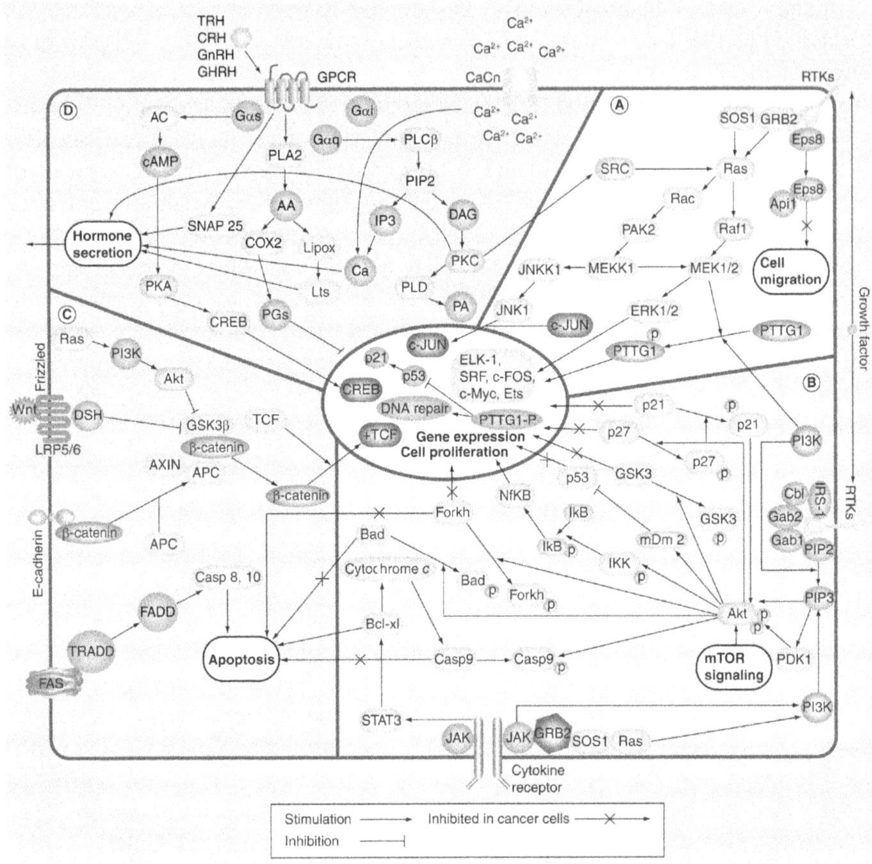

Figure 6.1 Main signaling pathways in pituitary tumorigenesis. (A) Growth factor-RTPKs pathway linked to cell migration via Eps8 or via Ras to the activation of oncogenes. (B) Growth factor-RTPKs pathway via PI3K and Akt network inhibits apoptosis. (C) Wnt/β-catenin pathway where Wnt family members bind to Frizzled receptors and induce TCF (T cell factor) expression. PTTG links several intracellular pathways regulating DNA repair and cell cycle via p53. (D) Hormones link to GPCR and mediated through Gα subunits can induce hormone secretion and CREB expression. APC, anaphase-promoting complex; CREB, cAMP response element binding; CRH, corticotropin-releasing hormone; GHRH, growth hormone-releasing hormone; GnRH, gonadotrophin-releasing hormone; GPCR, G-protein–coupled receptor; PTTG, pituitary tumor transforming gene; RTK, receptor tyrosine kinase; TRADD, TNF receptor-associated death domain; TRH, thyrotropin-releasing hormone.

It is obvious that although the hormone receptors are the same in normal and in neoplastic transformation in pituitary adenomas, the activation of different downstream substrates and targeted genes suggests that there are multiple levels of intracellular signaling complex interactions. We point out that hormones and specific receptor pathways use key signaling molecules that can be valuable to diagnostic endeavors. An example is the PI3K/AKT pathway, which is a possible target for somatostatin analogs [6].

Nuclear Receptors

Nuclear receptors are transcription factors that are activated upon binding to their ligands. Initially, they have been classified as classic endocrine nuclear hormone receptors and orphan receptors. This class of receptors includes peroxisome proliferators–activated receptors (PPARs) for fatty acids and several other receptor types. Other orphan receptors include retinoic acid receptor (RAR) for retinoid acids. PPAR and RAR were recently considered as potential therapeutic targets in ACTH-secreting tumors [35]. In most pituitary tumors, PPAR overexpression is associated with the absence of the transcriptional regulator COUP-TF1.

The most important nuclear hormone receptor is the ER, and its expression has been demonstrated in all types of pituitary tumors; the highest level was found in estrogen-responsive prolactinomas; significant ER levels are also present in GH/PRL and FSH/LH pituitary tumors. In addition, macroadenomas (especially male prolactinomas) are exhibiting a higher ER expression compared to microadenomas; this behavior explains the invasiveness of macroprolactinomas in men and provides support for the therapeutic approach for ER-based therapy. Estrogen-induced transcriptional targets include growth factors (bFGF, VEGF, IGF, EGF, and TGF) and oncogenic proteins (c-myc, cfos, erb, c-myb, and PTTG) [2].

Progesterone receptors (PRs) have also been mentioned in pituitary tumorigenesis. They have been proven to be expressed more frequently in pituitary adenomas found in women, and microadenomas show a higher expression of PR than macroadenomas. Progesterone is a potent regulator of GnRH and GnRHR I transcription [35]. Furthermore, PR agonists antagonize tumor development and functionality [36]. The role of these transcriptional regulators has not been clearly demonstrated yet, but they probably play a "permissive role" in pituitary tumor pathogenesis.

Signal Transduction Pathways

Although the number of signaling pathways may seem overwhelming, there is actually a relatively small number of distinct pathways for transducing external signals. The informational complexity needed to create many cell types and cell properties comes from combining these signals. Elucidation of the underlying principles and mechanisms relevant to all signaling pathways forms the foundation for understanding how cells integrate signals to achieve a particular identity or to develop a certain cellular response. External signals induce two major types of cellular responses: changes in the activity or function of specific preexisting proteins and changes in the amounts of specific proteins produced by a cell, most commonly as the result of transcription factor modification that leads to the modulation of gene transcription. In general, the first type of response occurs faster than the second type. Signaling from GPCRs often results in changes in the activity of preexisting proteins, although activation of these receptors on some cells can induce changes in gene expression as well [6].

PTTG Signaling

We are focusing on PTTG signaling because of its high involvement in normal physiological cellular events and in pituitary pathological transformation. PTTG is a proto-oncogene involved in pituitary and other tumors [37]. One of the functions of PTTG in facilitating pituitary tumor development is the induction of increased angiogenesis, a process mediated by factors like TGFβ3, bFGF, VEGF-A, VEGFR1, and VEGFR2, which appear to be associated with aggressive pituitary phenotype [38].

PTTG's normal physiological function is to act as a securin protein by inhibiting sister chromatid separation, thus regulating mitosis. Moreover, it plays a role in DNA repair, possibly by connecting DNA damage–response pathways with sister chromatid separation [39]. The key event in PTTG functioning is the nuclear translocation triggered by phosphorylation. This process may involve MAPK and PI3K pathways or CDK1, a dsDNA-dependent kinase. Regulation of PTTG expression by estrogens has been demonstrated. Most pituitary tumors express both high levels of ERs and PTTG; therefore, antiestrogen therapy may suppress PTTG expression and reduce tumor aggressiveness in the future [40]. Insulin and IGF1 were demonstrated to stimulate PTTG expression *in vitro* [41]. The PTTG promoter region contains insulin responsive motifs; insulin and IGF1 signaling cascades control PTTG expression and its translocation via MAPK and PI3K pathways.

PTTG is involved in chromosome stability by inhibiting ESPL1 (extra spindle pole bodies homolog 1) in mitosis, downregulating the p53 apoptotic pathway [42]. PTTG interacts with Sp1, which bridges binding to gene promoter regions. By suppressing p21 expression, PTTG interferes with p53 signaling. Upregulation of c-myc results from a direct interaction with PTTG1 [42]. Significant positive correlations between VEGF and PTTG mRNA expressions and between PTTG and KDR (kinase insert domain receptor) mRNA expressions were found in pituitary adenomas [43,44].

PTTG overexpression in pituitary tumors has been reported at the mRNA level for two decades. Northern blot analysis clearly differentiates tumoral from nontumoral pituitary. Immunostaining revealed that high cytoplasmic expression of PTTG in pituitary adenomas occurs very frequently (over 90%).

Several studies have investigated a potential correlation between PTTG expression and pituitary tumor subtypes and the potential predictive value of PTTG [37,39,43–45]. Cell type–dependent expression of PTTG was suggested by the significantly elevated expression found in GH-secreting adenomas compared to NF adenomas [37,39]. The level of expression, although somewhat higher in GH-secreting adenomas compared to PRL-, TSH-, or FSH-secreting adenomas, was not statistically significant; therefore, PTTG levels cannot differentiate secretory subtypes [37].

Minematsu et al. [44], examining pituitary adenomas, revealed elevated PTTG mRNA levels in most pituitary adenoma subtypes. In addition, PTTG-binding factor mRNA was elevated in all pituitary tumor subtypes, with significant correlation with PTTG expression [46].

The effect of PTTG on cell proliferation is unclear, as studies have reported contradictory findings. In its functional capacity as a securin, PTTG overexpression is expected to reduce cell proliferation by arrest of mitosis and inhibition of sister chromatid separation, whereas in its function as cell-transforming oncogene, one would expect a pro-proliferative action.

Correlation between PTTG expression and cell proliferation has been demonstrated in clinical studies [45]. For example, its expression correlated with proliferation marker immunopositivity, PCNA or Ki-67, in pituitary adenomas. The cutoff value separating recurring and nonrecurring tumors was 3.3% for PTTG immunopositivity. A cutoff of 2.9% for both PTTG and Ki-67 positivity predicted recurrence versus nonrecurrence, Ki-67 being the superior predictor [45]. Considering PRL-secreting pituitary tumors, elevated nuclear levels of PTTG were found in aggressive–invasive tumors in association with other histological features like elevated Ki-67 index, high mitotic index, and p53 abundance [40]. PTTG mRNA expression in hormone-secreting tumors is significantly higher in invasive tumors, while in NF tumors, PTTG levels did not correlate with tumor stage [39].

Some authors sustain the correlation between PTTG and angiogenic factors such as bFGF and VEGF. EGF induced PTTG1 expression and was associated with increased cell proliferation, elevated PCNA levels, and enhanced entry of cells into the S-phase [4]. Elevated levels of PTTG, FGF, and FGF-R1 were found in pituitary adenomas, yet the levels of all three molecules showed a weak correlation [46]. Correlation between PTTG and VEGF/VEGFR2 expressions was demonstrated [43], as well as colocalization of PTTG with VEGF [44]. Due to the correlation with angiogenic factors and the increased microvascular density, Minematsu et al. [44] suggested the implication of PTTG in angiogenic and cell proliferation functions.

Kanakis et al. [47] proposed that aberrant PTTG overexpression in pituitary adenomas may have other causes, including epigenetic factors such as hypomethylation, PTTG being involved in pituitary growth by paracrine mechanisms and the microenvironment. Growth factors are abundant within the pituitary microenvironment and may sustain the growth of pituitary adenomas by autocrine/paracrine feedback. EGF, which is known to act directly on hormone-secreting pituitary cells, is shown to affect cell proliferation and PTTG1 expression in hormonally inactive FSCs, the major source of pituitary cytokines and growth factors [48]. Pituitary adenomas could benefit from inhibition of EGFR-mediated PTTG1 expression in FSCs by intracellular blockade of paracrine signaling [49].

We can conclude that PTTG, being correlated with angiogenic factors, cell cycle components, and proliferation events, can be associated with tumor subtype, aggressiveness, and invasiveness and represents a good intracellular biomarker molecule.

Protein Kinases

Mitogen-Activated Protein Kinases

MAPKs are important players in signal transduction pathways. They are activated by a range of stimuli and mediate a number of physiological and pathological cellular functions. The MAPK pathway represents a cascade that links growth and differentiation signals with transcription. Growth factor receptors and tyrosine kinases activate Ras, which in turn activates c-Raf, MEK (mitogen-activated protein kinase), and MAPK. Activated p44/42 MAPK translocates to the nucleus and activates transcription by phosphorylation of kinases such as p90 RSK (ribosomal kinase), MSK (mitogen- and stress-activated protein kinase), and transcription factors such as ELK1 (E twenty-six (ETS)-like transcription factor) and STAT3 [50].

The ever-evolving MAPK pathways consist of four major groupings and numerous related proteins that constitute interrelated signal transduction cascades activated by a wide range of stimuli such as growth factors, stress, cytokines, and inflammation. The four major groupings are the ERK, JNK or SAPK (stress-activated protein kinase), p38, and the Big MAPK or ERK5 cascades. Signals from cell-surface receptors such as GPCRs and growth factor receptors are transduced, either directly or via small G-proteins such as Ras and Rac, to multiple protein kinases that amplify these signals and/or regulate each other.

The MAPK module intermediates signal transduction between short life-activated tyrosine-kinase receptors and downstream nuclear signaling. MAPK (ERK) activation is involved in the tumorigenic effects of TRH, GnRH, and GHRH [51]. In these cases, it does not play the role of a trigger, but of a bridge between primary events and downstream (nuclear) signaling. Thus, the MAPK pathway may be considered as a focal, high-sensitivity network for the detection of molecules that induce tumoral transformation and potential therapeutic targets because several inhibitors of this pathway showed significant potential in the treatment of different subtypes of pituitary tumors [1,21,52].

Protein Kinases A and C

Both the AKT and the MAPK pathways are overexpressed in many pituitary tumors, resulting in the inhibition of cell cycle inhibitors. These pathways share a common root in the tyrosine-kinase receptor; a change in these receptors or in their relationship to membrane matrix–related proteins may be an early event in tumorigenesis [1].

PKA, PKC, and PKB are critical for regulating the activity of many cellular proteins, often in response to external signals. All these protein kinases were found to be involved in pituitary transformation.

In the LβT2 mouse gonadotroph tumor cell line, it was demonstrated that activation of PKA is the major mechanism regulating the expression of the nuclear receptor Nur77 by GnRH, which may serve as a downstream signaling gene to mediate the antitumor effects of GnRH [53]. Overactivity of PKC pathway has been suggested in sporadic pituitary adenomas, and the PKCα isoform was found at high levels in a subset of aggressive pituitary tumors. In the case of pituitary somatotrophs, cAMP mediates a mitogenic signal; therefore, abnormally high cAMP generation leads to pituitary adenomas via PKA. High PKC expression in association with a PKC point mutation has been reported in few invasive pituitary tumors and PKC inhibitors abrogate pituitary tumor growth and induce apoptosis [2].

Phosphoinositol-3 Kinase/AKT

Many cell-surface receptors induce production of second messengers like PIP3 (phosphatidylinositol 3, 4, 5 trisphosphate), which conveys signals from the cellular surface to the cytoplasm. PIP3 signals activate the kinase PDK1 (3-phosphoinositide-dependent protein kinase-1) which in turn activates the kinase AKT, also known as protein kinase B (PKB). Proteins phosphorylated by activated AKT promote cell survival. Phosphorylation of Ikappa-B kinase leads to activation of the transcription factor NF-κB (Nuclear Factor-KappaB) to oppose apoptosis. Bad is a protein in the bcl-2 gene family that opposes bcl-2 to induce apoptosis. Phosphorylation of Bad by AKT blocks anti-apoptotic activity to promote cell survival. Similarly, phosphorylation via AKT of the protease caspase-9 or forkhead transcription factors can block the induction of apoptosis. AKT promotes cell survival and opposes apoptosis by a variety of routes.

Signaling through the PI3K family promotes cell proliferation and survival; thus, PI3K expression is strongly associated with transformation and metastases. Increased PI3K mRNA expression has recently been reported in pituitary tumors [2,54].

The abnormal activation of the PI3K/AKT pathway has been validated by epidemiological and experimental studies as an essential step toward the initiation and maintenance of human tumors. In mouse experimental models, it was demonstrated that activation of the PI3K/AKT pathway mediates, at least in part, the aberrant pituitary growth, and the intervention of this signaling pathway presents a novel therapeutic opportunity for TSHomas [55]. Activation of PI3K signaling promotes aberrant pituitary growth in a mouse model of TSH-secreting pituitary tumors. Constitutive pathway activation can result from the distinct and/or complementary biological events including (1) constitutively active mutants or amplification of RTKs, leading to constitutive recruitment and activation of PI3K and downstream effectors; (2) amplification of PI3K; (3) the presence of activating mutations in the PIK3CA gene encoding the p110a catalytic subunit; (4) overexpression of the downstream kinase AKT; (5) loss or inactivating mutations of the tumor suppressor gene PTEN (phosphatase and tensin homolog), an endogenous negative regulator of the PI3K pathway; or (6) constitutive recruitment and activation by mutant forms of the Ras oncogene [56].

In 2008, Lin et al. [57] showed strong genetic evidence supporting the role of the PI3K/AKT signaling pathway in the tumorigenesis of pituitary tumors, particularly in the invasive types. Therefore, relatively common PIK3CA mutations/amplifications, RAS mutations and their tendency of mutual exclusivity in pituitary tumors were reported.

Ras

Some classic oncogenes, such as *ras* and Myc, are involved in these endocrine tumors. Mutations in the *ras* GTP-binding or hydrolysis domains, corresponding to codons 12, 13, and 61, resulting in continuous *ras* activation are common early events in many solid tumors, including colorectal and thyroid cancer. However, *ras* activation is rare in pituitary tumors; *H-ras* mutations have been identified only in a single aggressive prolactinoma or metastases from pituitary carcinomas, not in the respective primary tumor [2].

In addition to the classic cancer genes, a significant number of genetic or epigenetic alterations in pituitary tumors target several cell cycle regulators as described in the following paragraphs. From these data, it has been estimated that more than 80% of pituitary tumors display alterations in at least one of the regulators of the G1/S transition of the cell cycle [58].

Wntβ-Catenin Pathways

Major pathways involved in the development of the pituitary include Notch and Wnt regulatory networks, which are mainly active in the early phases of pituitary organogenesis and are essential for the development of somatotropes, lactotropes, and thyrotropes [59].

Wnt signaling includes three different pathways: the canonical Wnt/β-catenin pathway, the noncanonical Wnt/c-JNK pathway, and the noncanonical Wnt/Ca2+ pathway. Aberrant activation of the Wnt/β-catenin signaling pathway is one of the most frequent abnormalities in human cancer.

It has recently been reported that Wnt proteins are overexpressed in various human cancers, but aberrant expression of Wnt proteins has not been described very extensively in human pituitary adenomas. A member of the Wnt family of genes, i.e., a growth factor, Wnt4 has been known to regulate proliferation of mouse anterior pituitary cell types during embryonic development [60].

However, the signaling pathways involving Wnt and its receptor in pituitary tumors are still a subject of controversy. Miyakoshi et al. [60] studied the expression of Wnt4 and its receptor, Frizzled 6 (Fzd6). They indicate that Wnt4 and Fzd6 were highly expressed in GHomas, PRLomas, and TSHomas, and prove that Wnt4/Fzd6 signaling was activated via the β-catenin-independent (noncanonical) pathway.

Several groups support the idea that Wnt4/Fzd6 signaling activates the canonical pathway. For example, Miyakoshi et al found β-catenin nuclear accumulation in the majority of investigated pituitary adenomas, demonstrating activation of the canonical Wnt/β-catenin pathway [60].

The β-catenin signaling pathway in pituitary tumorigenesis is still a subject of scientific debate. β-catenin is important in pituitary development because it controls the expression and transduction of a number of transcription factors and selected target genes. This signaling pathway also appears as a major regulator of the functioning pituitary through transcription/transduction of specific genes controlling hormone secretion, particularly of the gonadotroph lineage [61]. β-catenin plays an important but less critical role in pituitary tumor formation than it does in other types of more malignant tumors [61].

Activation of the canonical Wnt/β-catenin pathway, indicated by nuclear accumulation of β-catenin, is observed more often in other tumor types. In pituitary adenomas as well as in normal pituitaries, β-catenin membrane staining was detected by Elston et al. in 2008 [62]. However, cytoplasmic membrane accumulation of β-catenin has also been linked to carcinogenesis via E-cadherin. Components of the β-catenin pathway are detectable in benign pituitary tumors, this pathway being aberrantly activated, as reported in some studies. Pituitary adenomas show reduced expression of β-catenin and, in particular, of E-cadherin/catenin complexes in aggressive GHomas or PRLomas. In the same adenomas, the inactivation of E-cadherin by promoter hypermethylation also suggests that the alteration of cell–cell adhesion regulates the cytoplasmic pool of β-catenin and that both events play a role in the progression of these adenomas toward a more aggressive profile. Opposing findings showed overexpression of these two molecules in NFPAs and pituitary carcinomas [61,63].

FGFR and Wnt/β-catenin signaling pathways may interact and determine the invasive potential of pituitary tumors. FGFR4 and its truncated product, pdt-FGFR4, have been shown to play a crucial role in pituitary tumor invasion and progression, linking these processes to the cell–cell adhesion N-cadherin/β-catenin complexes and activation of Wnt/β-catenin signaling [61].

Wnt Inhibitory Factor-1

Wnt/β-catenin signaling can be inhibited by different endogenous antagonists (such as Wnt inhibitory factor-1 (WIF1), sFRP2, sFRP3, and sFRP4), among which WIF1 binds Wnt ligands, preventing them from binding to their receptors [62].

Elston et al. [62] have recently shown that a series of inhibitors of the Wnt pathway were all downregulated at the mRNA level in both functioning and NFPAs. Immunostaining showed underexpressed WIF1, suggesting that WIF1 may be a tumor suppressor, specifically in nonfunctioning pituitary tumors. Moreover, they associated cytoplasmic β-catenin staining with low values of WIF1 expression. This group also suggested that WIF1 mRNA downregulation in pituitary tumors is an early event because it is reduced in all tumor subtypes, and that later genetic events may explain the differences in tumor behavior seen in the various pituitary tumor subtypes.

This promoter inactivation or downregulation of the expression of Wnt inhibitors and the known upregulation of cyclin D1 expression suggests that aberrant activation of the canonical Wnt signaling pathway may also be involved in pituitary tumorigenesis [61].

In vitro studies have demonstrated the regulatory role of β-catenin on several transcription factors, switching them from a repressor function to a transcription activator function. β-catenin activation regulates the functioning of the somatotroph, mammotroph, thyrotroph, and gonadotroph pituitary axes, and it regulates pituitary-specific positive transcription factor 1 (Pit1) expression/transcription. The Wnt/β-catenin pathway also induces the transcription of the cell type–restricted transcription factor Pitx2, a LEF1 target gene that activates growth-regulatory genes (cyclin D1 and D2, and *c-Myc*) necessary for the proliferation of specific cells. Pitx2 itself modulates the amount of unstable Pitx2 mRNA induced by the activated canonical Wnt/β-catenin pathway and contributes to the stabilization of Pitx2 mRNA [61].

An interaction between the Wnt/β-catenin and the FGFR signaling pathways was reported at different levels: in particular, cross talk at the GSK3b node (which is inhibited through phosphorylation by the FGFR-activated PI3K/AKT pathway) links β-catenin and the SNAIL (Snail zinc finger protein) signaling cascades in the determination of cellular malignancy [62]. Thus, SNAIL represses the transcription of the CDH1 gene (which encodes E-cadherin) and therefore regulates the cytoplasmic pool of β-catenin, enhancing Wnt-induced β-catenin translocation to the nucleus [61].

The Wnt/β-catenin pathway comprises promising signaling molecules as pituitary diagnostic tools that especially detect the aggressiveness of the tumor, while Wnt inhibitors might prove to be valuable in a panel of signaling molecules used for subtyping pituitary adenomas.

Transcription Factors

Pit1 is a transcription factor involved in the specification of the lactotrope, somatotrope, and thyrotrope phenotypes in the developing anterior pituitary. In somatotroph tumors, the downregulation of Pit1 is mediated by the PI3/AKT pathway and governs cell survival. GHRH is able to activate the PI3/AKT pathway and regulate Pit1 transcription [64].

The presence of pituitary homeobox factor 1 (Ptx1), described as a POMC (Pro-opiomelanocortin) gene expression activator, has been demonstrated in all normal anterior pituitary cells and in most pituitary adenoma subtypes. Reduced expression of Ptx1 was found in pituitary tumors, while expression of the transactivating factor Prop1 is not affected. Ptx2 expression was demonstrated in pure lactotroph tumors, suggesting that this transcription factor plays a critical role in the differentiation of secretory cell phenotypes.

Summing up, in intracellular signaling pathways, a major molecule involved in pituitary tumorigenesis is PTTG, which is linked in an intricate network with hormones and growth factors. Increasing evidence supports a multifunctional role of PTTG1 in cell physiology and tumorigenesis. Tumorigenic mechanisms for PTTG1 action involve cell transformation, aneuploidy, apoptosis, angiogenesis, and tumorigenic microenvironment feedback. Owing to its central role, it can definitely be a diagnosis marker. Moreover, various important roles in the continuous quest for key signaling molecules was demonstrated for Wnt/β-catenin pathway components.

Cell Cycle

The cell cycle is regulated by the interplay of many molecules. Key players among these are the cyclins, which are expressed and then degraded in a concerted fashion to drive the cell cycle phases. Cyclins combine with CDKs to form activated kinases that phosphorylate targets, leading to cell cycle regulation. A breakdown in the regulation of this cycle can lead to uncontrolled growth and tumor formation. Defects in many of the molecules that regulate the cell cycle have been implicated in cancer. Important molecules among these are p53, CDK inhibitors (such as p15, p16, p18, p19, p21, and p27), and Rb. All of these act to keep the cell cycle from progressing until all damaged DNA has been repaired.

From the up-to-date literature results, among the molecules involved in the cell cycle signaling pathway, cyclins and retinoblastoma proteins are relevant to pituitary adenomas [58]. Among various kinases, CDKs and their inhibitors are critical regulators of the transition through different cell cycle phases [65]. The primary substrates of the CDKs in G1 progression are the members of the retinoblastoma protein family (pRb) that negatively regulate entry into the cell cycle and G1/S progression [58].

Cyclins

Immunocytochemistry for cyclin D1 protein expression failed to demonstrate its increased expression, in the context of a described allelic imbalance of cell cycle nuclear D1 gene in one-quarter of pituitary tumors. The significance of this finding, thus, is unclear [2]. Methylation in exons for CDK inhibitors, such as p16, was reported in the majority of pituitary tumors. Another study demonstrated p16/*CDKN2A* gene methylation in the majority of NF tumors and in 9.5% of somatotrophinomas [2]. Genetic ablation of both p16INK4a and p18INK4c cooperates in the incidence and the latency of pituitary tumor development [66]. The former models suggest a clear relevance of the CDK/pRb pathway (and their inhibitors INK4 or KIP) in pituitary tumorigenesis [58].

Retinoblastoma-Associated Protein

In experimental models, tumor incidence provoked by the partial deletion of pRb is partially reverted by a mutation in pRb effectors such as E2f1 or E2f4, indicating the relevance of the pRb/E2F pathway in pituitary tumorigenesis [58]. Several studies based on immunodetection in tumor sections found abnormal expression of pRb in different pituitary adenomas. In some cases, decreased expression correlated with an epigenetic mechanism, namely hypermethylation of the pRb promoter [2,67] or deletion within the protein-pocket-binding domain [58]. The protein was demonstrated to be linked with PTTG. In the pituitary of Rb-transgenic mice, overexpression of PTTG1 generates pituitary hyperplasia, focal microadenomas, and higher incidence of tumors. Losses in the chromosomal region 13g14.2, where tumor suppressor rb gene resides, are related to aggressive human pituitary tumor behavior and the lack of expression of the protein observed in one-fourth of GH-secreting pituitary adenomas [68].

Protein 27

Protein 27 (p27) regulates G1 cell cycle progression, and low expression has been reported in pituitary adenomas in comparison with normal pituitary tissue [2]. Recent data suggest the existence of two major pathways for G1/S-phase deregulation in pituitary tumors. One branch is formed of p18INK4c/CAKE/pRb, whereas the other one is represented by p27Kip1 and perhaps p21Cip1 [58].

Based on the cell cycle array of effectors, we can conclude that in the pituitary transformation, CDK4/pRb pathways are likely to provide valuable information in diagnosis and/or provide therapeutic targets. A brief synthesis on relevant signaling molecules involved in pituitary tumors is presented in Table 6.1.

Pituitary tumorigenesis still has uncovered aspects, and we are just beginning to unveil some of the components of a complicated maze. There is cross talk between microenvironmental factors and pituicytes that leads to tumor transformation, a process that can be modulated at various levels.

As pituitary tumors are characterized by overactivity of AKT and MAPK pathways, they are definitively targets for inhibition mediated by somatostatin analogs.

Table 6.1 Signaling Molecules Involved in Pituitary Tumors

Molecule	Gene Location	UniprotKB Code	Recommended and Alternative Names	Biological Effects and Modulation	Status in Pathology and Genetics	References
AIP	11q13.3	O00170	Aryl hydrocarbon receptor-interacting protein (AIP), XAP2; Immunophilin homolog ARA9; hepatitis B virus X-associated protein 2	Modulator of aryl hydrocarbon receptor receptivity for ligand and/or its nuclear targeting	Defects in AIP cause pituitary adenoma predisposition, familial isolated pituitary adenoma, GH- and ACTH-secreting pituitary adenoma Mutations relevant to pituitary adenoma: • 16 R→H in familial isolated pituitary adenoma and pituitary adenoma predisposition • Deletion of 248 Tyr (ACTH-secreting pituitary adenoma) • 304 R→Q (ACTH-secreting pituitary adenoma)	[69–72]
AKT	14q32.3 8p21.1	P31749	PKB AKT1-human	Mediating effects of growth factors (PDGF (platelet-derived growth factor), EGF), IGF-I, and PI3K	Overexpressed in pituitary adenoma Enhanced signaling via AKT pathway in pituitary adenoma	[1]
bFGF	4q26-q27	P09038	Heparin-binding growth factor 2; bFGF, FGF2	Angiogenic, mitogen, interacts with CSPG4 and FGFBP1	Correlates with tumor size and aggression	[73]
CDH1	16q22.1	P12830	E-cadherin, cadherin-1; uvomorulin, CAM (Calmodulin) 120/80, CD324	Involved in cell-adhesion, potent invasive suppressor role	Defects in *CDH1* are involved in dysfunction of the cell–cell adhesion system, triggering cancer invasion and metastasis	[74]

CDK1	10q21.1	P06493	Cell division control protein 2 homolog; p34 protein kinase	Controls cell cycle, required for entry into S-phase and mitosis. Part of the kinase complex phosphorylating RNA polymerase 2	Involved in PTTG nuclear translocation	[75]
CDK1				Inactivated by Thr-14 or Tyr-15 phosphorylation; activated by Thr-161 phosphorylation		
CDK4	12q14	P11802	Cell division protein kinase 4; CDK4; PSK-J3	Control of cell cycle progression Thr-172 phosphorylation essential for enzymatic activity	Possibly involved in deregulation of G1/S-phase	[76]
c-myb	6q22-23	P10242	Myb proto-oncogene protein	Transcriptional activator; DNA-binding protein Binds *MYBBP1A*. Interacts with homeodomain-interacting protein kinase 2, macrophage activating factor, and nemo-like kinase	Overexpressed in prolactinomas owing to ER stimulation	[77]
c-myc	8q24.12-13	P01106	Myc proto-oncogene protein; transcription factor p64	Transcription regulator for growth-related genes Phosphorylation by DNA-dependent protein kinase catalytic subunit	Overexpression of MYC is implicated in the multistep tumorigenesis of pituitary tumors	[78]

(*Continued*)

Table 6.1 (Continued)

Molecule	Gene Location	UniprotKB Code	Recommended and Alternative Names	Biological Effects and Modulation	Status in Pathology and Genetics	References
COUP-TF1	5q14	P10589	COUP transcription factor 1 Nuclear receptor subfamily 2 group F member 1	Nuclear steroid receptor In conjunction with S300-II stimulates initiation of transcription	Downregulated/absent in pituitary adenoma	[35,79]
COX-2	1q25.2-q25.3	P00403	Prostaglandin G/H synthase 2; COX2; prostaglandin-endoperoxide synthase 2; prostaglandin H2 synthase 2	Major mediator of inflammation and/or a role for prostanoid signaling in activity-dependent plasticity	Defects in *MT-CO2* are associated with tumor formation	[80]
CREB	2q34	P16220	CREB protein 1	Phosphorylation of Ser-133 and Ser-142 activates CREB	Enhanced Ser-133-phosphorylation in pituitary adenoma	[2]
Cyclin D1	11q13	P24385	G1/S-specific cyclin D1 PRAD1 oncogene CCND1	Cell cycle control	Upregulated in pituitary adenoma	[61]
Eps8	12q23-24	Q12929	EGFR kinase substrate 8	Upon binding to EGFR enhances EGF-dependent mitogenic signals. Can bind multiple cellular targets Phosphorylated by several RTKs	Overexpressed in pituitary adenoma	[5]

ErbB2	17q11.2-q12	P04626	Receptor tyrosine-protein kinase erbB2	Positive regulation of MAPK activity	Highly expressed in GH, PRL, or GH/PRL secretory pituitary adenoma	[81]
ESPL1	12q13	Q14674	Separin	Chromosome segregation Inactivated by interaction with securin/PTTG1 Inactivated by Ser-1126-phosphorylation Activated by removal of inhibitors	Altered activity in pituitary adenoma	[82]
FGF4	11q13.3	P08620	FGF4; heparin secretory-transforming protein; heparin-binding growth factor 4	Mitogenic activity	Highly expressed in pituitary adenoma	[2]
Fzd6	8q22.3-23.1	O60353	Fzd6	Receptor for Wnt proteins	Highly expressed in secretory pituitary adenoma (GH, PRL, TSH)	[60]
GSK3b	3p21	P35222	Catenin β-1	Wnt signaling pathway component Mediator in hormonal control of MYB and JUN Inhibited when phosphorylated by AKT1		[83]

(Continued)

Table 6.1 (Continued)

Molecule	Gene Location	UniprotKB Code	Recommended and Alternative Names	Biological Effects and Modulation	Status in Pathology and Genetics	References
HIF	1q21	Q16665	HIF	Transcriptional regulator of the adaptive response to hypoxia Upregulated in hypoxia Receptor-mediated, induced extracellular modulators: PDGF, EGF, FGF-2, IGF-2, TGF-1β, HGF (hepatocyte growth factor), TNFα, IL-1β (interleukin), angiotensin-2, and thrombin	Overexpressed in most common cancers and their metastases because of mutations in genes encoding oncoproteins and tumor suppressors	[84]
INK4	9p21	P42771	CDK4 inhibitor A; p16	Negative regulator of proliferation of normal cells; inhibits CDK4 and CDK6 interaction with cyclin D and phosphorylation of retinoblastoma protein	Defects in *CDKN2A* are involved in tumor formation in a wide range of tissues Inactivated in pituitary adenoma	[85]
JNK	10q11.22	P45983	MAPK-8 c-JNK	Regulator of AP-1 transcriptional activity Activated by Thr/Tyr phosphorylation by MAP2K4 and MAP2K7 Inactivated by dual specificity phosphatases (DUSP1)	Overactivated in pituitary adenoma in response to hormones and/or growth factors	[24–28]

Notch3	19p13.1-2	Q9UM47	Neurogenic locus notch homolog protein 3	Functions as a receptor for membrane-bound ligands Jagged1, Jagged2, and Delta1 to regulate cell-fate determination	Affects differentiation, proliferation, and apoptotic processes	[59]
p21Cip1	6p21.2	P38936	CDK inhibitor 1, P21, CAP20, CDKN1	Inhibits CDK activity Phosphorylation of Thr-145 by AKT or of Ser-146 by PKC	Involved in G1/S-phase deregulation in pituitary tumors	[58]
p27Kip1	12p12-13.1	P46527	CDK inhibitor 1B CDK inhibitor p27	Important regulator of cell cycle progression involved in G1 arrest Potent inhibitor of cyclin E- and cyclin A-CDK2 complexes Positive regulator of CDK4 Inactivation by phosphorylation and degradation	Defects in *CDKN1B* are the cause of multiple endocrine neoplasia type 4	[86]
Pit1	3p11	P28069	Pituitary-specific positive transcription factor 1	Control of cell proliferation Transcription factor involved in the lactotrope, somatotrope, and thyrotrope phenotypes Activates growth hormone and prolactin genes	Defects in *POU1F1* induce familial combined pituitary hormone deficiency, with impaired production of GH and one or more of the other five anterior pituitary hormones	[78]

(*Continued*)

Table 6.1 (Continued)

Molecule	Gene Location	UniprotKB Code	Recommended and Alternative Names	Biological Effects and Modulation	Status in Pathology and Genetics	References
Pitx2	4q25-27	Q99697	Pituitary homeobox 2	Differentiation of secretory cell types	Present in secretory pituitary adenoma	[87]
PRB	20q11.2	P28749	Retinoblastoma-associated protein, PRB1, retinoblastoma-like protein 1	Inhibitor of E2F-mediated transactivation Regulator of entry into cell division Cell cycle arrest properties inactivated by CDK4 through phosphorylation on Thr-332, Ser-640, Ser-964, and Ser-975	Mutations: 640 S→A: Strongly reduces phosphorylation by CDK2 and CDK4 657–660 KRRL→AAAA: Reduces S-640 phosphorylation by CDK2 and CDK4	[88]
Prop-1	5q35.3	O75360	Homeobox protein prophet of Pit1	Ontogenesis of pituitary gonadotropes, somatotropes, lactotropes, and thyrotropes	Defects in *PROP1* induce combined pituitary hormone deficiency	[87]
PTEN	10q23.3	P60484	Phosphatase and tensin homolog, phosphatidylino-sitol 3, 4, 5 trisphosphate 3-phosphatase	Tumor suppressor by negatively regulating AKT/PKB pathway	Mutations of *PTEN* reported in many types of tumors, including endocrine tumors	[89]

PTTG1	5q35.1	O95997	Securin	Regulator of chromosome stability	Aneuploidy due to *PTTG1* overexpression in different cancers, including pituitary	[37,39, 43–45]
			PTTG1 protein	Negatively regulates the transcriptional activity and related apoptosis activity of TP53		
Ptx1	5q31	P78337	Paired-like homeodomain 1, pituitary homeobox 1, Pitx1	Involved in intracellular signaling, triggering GnRH response	Reduced level in pituitary adenoma	[87]
Ras	11p15.5	P01112	GTPase HRas p21ras	Ras proteins bind GDP/GTP and possess intrinsic GTPase activity	Overstimulated in pituitary adenoma Mutations in positions 12, 13, or 61 implicated in human tumors	[66,89]
RECK	9p12-13	O95980	RECK	Suppressor of MMP9 secretion and inhibitor of its enzymatic activity	Downregulation by oncogenic signals may facilitate tumor invasion and metastasis	[3]
			Suppressor of tumorigenicity protein 15 (ST 15)	Also regulates MMP2 and MT1-MMP		
SNAIL	20q13.1-2	O95863	Zinc finger protein SNAIL; protein snail homolog 1	Binds to 3 E-boxes of the E-cadherin gene promoter and represses its transcription		[61]
WIF1	12q13.13	Q9Y5W5	Wnt inhibitory factor 1	Binds to Wnt proteins and inhibits their activities	Downregulated in pituitary adenoma	[62]
Wnt4	1p36.1	P56705	Protein Wnt-4	Ligand for members of the frizzled family of seven TM receptors	Highly expressed in secretory pituitary adenomas—GH, PRL, TSH	[60]

ERK1/2 is chronically activated after both gsp oncogene expression and Gsα overexpression. Moreover, activation of ERK1/2 is required to observe hormonal hypersecretion induced by Gsα alterations.

Thus, besides the setting up of the compensatory mechanisms of cAMP pathway activation, as identified in gsp+ tumors, Gsα overexpression and the gsp oncogene might impact the tumoral phenotype by targeting the ERK1/2 pathway. Increasing lines of evidence suggest clinically significant applications of PTTG1 in correlation with aggressive phenotype or survival rate, and classifies it as an interesting candidate for malignancy, tumor staging, and as a biomarker for subsequent therapeutic interventions.

The Wnt/β-catenin pathway offers promising signaling molecules for diagnosis, detecting the aggressiveness of the tumor, while Wnt inhibitors can contribute to the panel of signaling molecules for pituitary adenoma subtyping. Future research should concentrate on understanding the molecular mechanisms that govern pituitary tumor transformation, where intracellular signaling molecules will constitute not only diagnostic/prognostic markers, but also novel therapeutical targets.

References

1. Grossman, A.B. (2009) The molecular biology of pituitary tumours: A personal perspective. *Pituitary*, 12(3): 265–70.
2. Heaney, A.P. (2005) Pituitary tumour pathogenesis. *Br. Med. Bull.*, 75–76: 81–97.
3. Yoshida, D., Nomura, R. and Teramoto, A. (2008) Regulation of cell invasion and signalling pathways in the pituitary *adenoma* cell line, HP-75, by reversion-inducing cysteine-rich protein with kazal motifs (RECK). *J. Neurooncol.*, 89(2): 141–50.
4. Vlotides, G., Eigler, T. and Melmed, S. (2007) Pituitary tumor-transforming gene: Physiology and implications for tumorigenesis. *Endocr. Rev.*, 28(2): 165–86.
5. Xu, M., Shorts-Cary, L., Knox, A.J., Kleinsmidt-Demasters, B., Lillehei, K. and Wierman, M.E. (2009) Epidermal growth factor receptor pathway substrate 8 (Eps8) is over expressed in human pituitary tumours: Role in proliferation and survival. *Endocrinology*, 150(5): 2064–71.
6. Tanase, C.P., Neagu, M. and Albulescu, R. (2009) Key signaling molecules in pituitary tumors. *Expert. Rev. Mol. Diagn.*, 9(8): 859–77.
7. Pertuit, M., Barlier, A., Enjalbert, A. and Gérard, C. (2009) Signalling pathway alterations in pituitary adenomas: Involvement of Gsα, cAMP and mitogen-activated protein kinases. *J. Neuroendocrinol.*, 21(11): 869–77.
8. Picard, C., Silvy, M., Gerard, C., Buffat, C., Lavaque, E., Figarella-Branger, D., et al. (2007) Gsα overexpression and loss of Gsα imprinting in human somatotroph adenomas: Association with tumor size and response to pharmacologic treatment. *Int. J. Cancer*, 121: 1245–52.
9. Naviglio, S., Pagano, M., Romano, M., Sorrentino, A., Fusco, A., Illiano, F., et al. (2004) Adenylate cyclase regulation via proteasome-mediated modulation of Gαs levels. *Cell Signal.*, 16: 1229–37.
10. Weinstein, L.S., Liu, J., Sakamoto, A., Xie, T. and Chen, M. (2004) Minireview: GNAS: Normal and abnormal functions. *Endocrinology*, 145: 5459–64.
11. Weinstein, L.S., Chen, M., Xie, T. and Liu, J. (2006) Genetic diseases associated with heterotrimeric G proteins. *Trends Pharmacol. Sci.*, 27: 260–66.

12. Hayward, B.E., Barlier, A., Korbonits, M., Grossman, A.B., Jacquet, P., Enjalbert, A., et al. (2001) Imprinting of the G(s)α gene GNAS1 in the pathogenesis of acromegaly. *J. Clin. Invest.*, 107: R31–36.
13. Lugnier, C. (2006) Cyclic nucleotide phosphodiesterase (PDE) superfamily: A new target for the development of specific therapeutic agents. *Pharmacol. Ther.*, 109: 366–98.
14. Persani, L., Borgato, S., Lania, A., Filopanti, M., Mantovani, G., Conti, M., et al. (2001) Relevant cAMP-specific phosphodiesterase isoforms in human pituitary: Effect of Gs(α) mutations. *J. Clin. Endocrinol. Metab.*, 86: 3795–800.
15. Peri, A., Conforti, B., Baglioni-Peri, S., Luciani, P., Cioppi, F., Buci, L., et al. (2001) Expression of cyclic adenosine 3′,5′-monophosphate (cAMP)-responsive element binding protein and inducible cAMP early repressor genes in growth hormone-secreting pituitary adenomas with or without mutations of the Gsα gene. *J. Clin. Endocrinol. Metab.*, 86: 2111–17.
16. Romano, D., Magalon, K., Pertuit, M., Rasolonjanahary, R., Barlier, A., Enjalbert, A., et al. (2007) Conditional overexpression of the wild-type Gsα as the gsp oncogene initiates chronic extracellularly regulated kinase 1/2 activation and hormone hypersecretion in pituitary cell lines. *Endocrinology*, 148: 2973–83.
17. Sakai, N., Kim, K., Sanno, N., Yoshida, D., Teramoto, A. and Shibasaki, T. (2008) Elevation of growth hormone-releasing hormone receptor messenger ribonucleic acid expression in growth hormone-secreting pituitary adenoma with Gsα protein mutation. *Neurol. Med. Chir.*, 48(11): 481–87.
18. Asa, S.L., Digiovanni, R., Jiang, J., Ward, M.L., Loesch, K., Yamada, S., et al. (2007) A growth hormone receptor mutation impairs growth hormone autofeedback signalling in pituitary tumours. *Cancer Res.*, 67(15): 7505–11.
19. Kontogeorgos, G. and Kovacs, K. (2004) *Molecular Pathology of the Pituitary*. Karger, Basel.
20. Theodoropoulou, M., Zhang, J., Laupheimer, S., Paez-Pereda, M., Erneux, C., Florio, T., et al. (2006) Octreotide, a somatostatin analogue, mediates its antiproliferative action in pituitary tumor cells by altering phosphatidylinositol 3-kinase signalling and inducing Zac1 expression. *Cancer Res.*, 66(3): 1576–82.
21. Hubina, E., Nanzer, A.M., Hanson, M.R., Ciccarelli, E., Losa, M., Gaia, D., et al. (2006) Somatostatin analogues stimulate p27 expression and inhibit the MAP kinase pathway in pituitary tumours. *Eur. J. Endocrinol.*, 155(2): 371–79.
22. Ferone, D., Gatto, F., Arvigo, M., Resmini, E., Boschetti, M., Teti, C., et al. (2009) The clinical-molecular interface of somatostatin, dopamine and their receptors in pituitary pathophysiology. *J. Mol. Endocrinol.*, 42(5): 361–70.
23. Naor, Z. (2009) Signalling by G-protein–coupled receptor (GPCR): Studies on the GnRH receptor. *Front. Neuroendocrinol.*, 30(1): 10–29.
24. Dobkin-Bekman, M., Naidich, M., Pawson, A.J., Millar, R.P., Seger, R. and Naor, Z. (2006) Activation of mitogen-activated protein kinase (MAPK) by GnRH is cell-context dependent. *Mol. Cell Endocrinol.*, 252(1–2): 184–90.
25. Kraus, S., Levy, G., Hanoch, T., Naor, Z. and Seger, R. (2004) Gonadotropin-releasing hormone induces apoptosis of prostate cancer cells: Role of c-Jun NH2-terminal kinase, protein kinase B, and extracellular signal-regulated kinase pathways. *Cancer Res.*, 64(16): 5736–44.
26. Kraus, S., Naor, Z. and Seger, R. (2006) Gonadotropin-releasing hormone in apoptosis of prostate cancer cells. *Cancer Lett.*, 234(2): 109–23.
27. Maudsley, S., Davidson, L., Pawson, A.J., Chan, R., Lopez de Maturana, R. and Millar, R.P. (2004) Gonadotropin-releasing hormone (GnRH) antagonists promote proapoptotic

signalling in peripheral reproductive tumor cells by activating a Gαi-coupling state of the type I GnRH receptor. *Cancer Res.*, 64(20): 7533–44.
28. Kim, K.Y., Choi, K.C., Park, S.H., Auersperg, N. and Leung, P.C. (2005) Extracellular signal-regulated protein kinase, but not c-Jun N-terminal kinase, is activated by type II gonadotropin-releasing hormone involved in the inhibition of ovarian cancer cell proliferation. *J. Clin. Endocrinol. Metab.*, 90(3): 1670–77.
29. Massa, A., Casagrande, S., Bajetto, A., Porcile, C., Barbieri, F., Thellung, S., et al. (2006) SDF-1 controls pituitary cell proliferation through the activation of ERK1/2 and the Ca2+ -dependent, cytosolic tyrosine kinase Pyk2. *Ann. N. Y. Acad. Sci.*, 1090: 385–98.
30. Liu, W., Matsumoto, Y., Okada, M., Miyake, K., Kunishio, K., Kawai, N., et al. (2005) Matrix metalloproteinase 2 and 9 expression correlated with cavernous sinus invasion of pituitary adenomas. *J. Med. Invest.*, 52(3–4): 151–58.
31. Gong, J., Zhao, Y., Abdel-Fattah, R., Amos, S., Xiao, A., Lopes, M.B., et al. (2008) Matrix metalloproteinase-9, a potential biological marker in invasive pituitary adenomas. *Pituitary*, 11(1): 37–48.
32. Hussaini, I.M., Trotter, C., Zhao, Y., Abdel-Fattah, R., Amos, S., Xiao, A., et al. (2007) Matrix metalloproteinase-9 is differentially expressed in nonfunctioning invasive and noninvasive pituitary adenomas and increases invasion in human pituitary adenoma cell line. *Am. J. Pathol.*, 170(1): 356–65.
33. Hazell, G., Yao, S., Roper, J., Prossnitz, E., O'Carroll, A.M. and Lolait, S. (2009) Localisation of GPR30, a novel G-protein–coupled oestrogen receptor, suggests multiple functions in rodent brain and peripheral tissues. *J. Endocrinol.*, 202(2): 223–36.
34. Sakamoto, H., Matsuda, K., Hosokawa, K., Nishi, M., Morris, J.F., Prossnitz, E.R., et al. (2007) Expression of G-protein–coupled receptor-30, a G-protein–coupled membrane estrogen receptor, in oxytocin neurons of the rat paraventricular and supraoptic nuclei. *Endocrinology*, 148(12): 5842–50.
35. Heaney, A.P., Fernando, M., Yong, W.H. and Melmed, S. (2002) Functional PPAR-gamma receptor is a novel therapeutic target for ACTH-secreting pituitary adenomas. *Nat. Med.*, 8(11): 1281–87.
36. Rey-Roldan, E.B., Grillo, C.A., Pietranera, L., Libertun, C., Nicola, A.F. and Piroli, G.G. (2008) Levonorgestrel antagonism on estrogen-induced pituitary tumours is mediated by progesterone receptors. *Horm. Metab. Res.*, 40(4): 245–50.
37. Hunter, J.A., Skelly, R.H., Aylwin, S.J., Geddes, J.F., Evanson, J., Besser, G.M., et al. (2003) The relationship between pituitary tumour transforming gene (PTTG) expression and *in vitro* hormone and vascular endothelial growth factor (VEGF) secretion from human pituitary adenomas. *Eur. J. Endocrinol.*, 148(2): 203–11.
38. Boikos, S.A. and Stratakis, C.A. (2007) Molecular genetics of the cAMP-dependent protein kinase pathway and of sporadic pituitary tumorigenesis. *Hum. Mol. Genet.*, 16(1): R80–87.
39. Salehi, F., Kovacs, K., Scheithaur, B.W., Lloyd, R.V. and Cusimano, M. (2008) Pituitary tumor-transforming gene in endocrine and other neoplasms: A review and update. *Endocr. Relat. Cancer*, 15(3): 721–43.
40. Wierinckx, A., Auger, C., Devauchelle, P., Reynaud, A., Chevallier, P., Jan, M., et al. (2007) A diagnostic marker set for invasion, proliferation, and aggressiveness of prolactin pituitary tumours. *Endocr. Relat. Cancer*, 14(3): 887–900.
41. Chamaon, K., Kirches, E., Kanakis, D., Braeuninger, S., Dietzmann, K. and Mawrin, C. (2005) Regulation of the pituitary tumor transforming gene by insulin-like-growth factor-I and insulin differs between malignant and non-neoplastic astrocytes. *Biochem. Biophys. Res. Commun.*, 331(1): 86–92.

42. Hamid, T. and Kakar, S.S. (2004) PTTG/securin activates expression of p53 and modulates its function. *Mol. Cancer*, 3: 18.
43. McCabe, C.J., Boelaert, K., Tannahill, L.A., Heaney, A.P., Stratford, A.L., Khaira, J.S., et al. (2002) Vascular endothelial growth factor, its receptor KDR/Flk-1, and pituitary tumor transforming gene in pituitary tumours. *J. Clin. Endocrinol. Metab.*, 87(9): 4238–44.
44. Minematsu, T., Suzuki, M., Sanno, N., Takekoshi, S., Teramoto, A. and Osamura, R.Y. (2006) PTTG over expression is correlated with angiogenesis in human pituitary adenomas. *Endocr. Pathol.*, 17(2): 143–53.
45. Filippella, M., Galland, F., Kujas, M., Young, J., Faggiano, A., Lombardi, G., et al. (2006) Pituitary tumour transforming gene (PTTG) expression correlates with the proliferative activity and recurrence status of pituitary adenomas: A clinical and immunohistochemical study. *Clin. Endocrinol. (Oxf.)*, 65(4): 536–43.
46. McCabe, C.J., Khaira, J.S., Boelaert, K., Heaney, A.P., Tannahill, L.A., Hussain, S., et al. (2003) Expression of pituitary tumour transforming gene (PTTG) and fibroblast growth factor-2 (FGF-2) in human pituitary adenomas: Relationships to clinical tumour behaviour. *Clin. Endocrinol. (Oxf.)*, 58(2): 141–50.
47. Kanakis, D., Kirches, E., Mawrin, C. and Dietzmann, K. (2003) Promoter mutations are no major cause of PTTG over expression in pituitary adenomas. *Clin. Endocrinol. (Oxf.)*, 58(2): 151–55.
48. Vlotides, G., Zitzmann, K., Hengge, S., Engelhardt, D., Stalla, G.K. and Auernhammer, C.J. (2004) Expression of novel neurotrophin-1/B-cell stimulating factor-3 (NNT-1/BSF-3) in murine pituitary folliculostellate TtT/GF cells: Pituitary adenylate cyclase-activating polypeptide and vasoactive intestinal peptide-induced stimulation of NNT-1/BSF-3 is mediated by protein kinase A, protein kinase C, and extracellular-signal-regulated kinase 1/2 pathways. *Endocrinology*, 145(2): 716–27.
49. Vlotides, G., Cruz-Soto, M., Rubinek, T., Eigler, T., Auernhammer, C.J. and Melmed, S. (2006) Mechanisms for growth factor-induced pituitary tumor transforming gene-1 expression in pituitary folliculostellate TtT/GF cells. *Mol. Endocrinol.*, 20(12): 3321–35.
50. Friday, B.B. and Adjei, A.A. (2008) Advances in targeting the Ras/Raf/MEK/Erk mitogen-activated protein kinase cascade with MEK inhibitors for cancer therapy. *Clin. Cancer Res.*, 14(2): 342–46.
51. Fougner, S.L., Bollerslev, J., Latif, F., Hald, J.K., Lund, T., Ramm-Pettersen, J., et al. (2008) Low levels of raf kinase inhibitory protein in growth hormone-secreting pituitary adenomas correlate with poor response to octreotide treatment. *J. Clin. Endocrinol. Metab.*, 93(4): 1211–16.
52. Jackson, T.A., Koterwas, D.M., Morgan, M.A. and Bradford, A.P. (2003) Fibroblast growth factors regulate prolactin transcription via an atypical Rac-dependent signalling pathway. *Mol. Endocrinol.*, 17(10): 1921–30.
53. Hamid, T., Malik, M.T., Millar, R.P. and Kakar, S.S. (2008) Protein kinase A serves as a primary pathway in activation of Nur77 expression by gonadotropin-releasing hormone in the Lβ T2 mouse pituitary gonadotroph tumor cell line. *Int. J. Oncol.*, 33(5): 1055–64.
54. Grossman, A.B. and Korbonits, M. (2004) Akting and cycling: A tale of the pituitary. *Horm. Res.*, 62(Suppl. 3): 117–23.
55. Lu, C., Willingham, M.C., Furuya, F. and Cheng, S.Y. (2008) Activation of phosphatidylinositol 3-kinase signalling promotes aberrant pituitary growth in a mouse model of thyroid-stimulating hormone-secreting pituitary tumours. *Endocrinology*, 149(7): 3339–45.
56. Garcia-Echeverria, C. and Sellers, W.R. (2008) Drug discovery approaches targeting the PI3K/AKT pathway in cancer. *Oncogene*, 27(41): 5511–26.

57. Lin, Y., Jiang, X., Shen, Y., Li, M., Ma, H., Xing, M., et al. (2009) Frequent mutations and amplifications of the PIK3CA gene in pituitary tumours. *Endocr. Relat. Cancer*, 16(1): 301–10.
58. Quereda, V. and Malumbres, M. (2009) Cell cycle control of pituitary development and disease. *J. Mol. Endocrinol.*, 42(2): 75–86.
59. Zhu, X., Wang, J., Ju, B.G. and Rosenfeld, M.G. (2007) Signalling and epigenetic regulation of pituitary development. *Curr. Opin. Cell Biol.*, 19(6): 605–11.
60. Miyakoshi, T., Takei, M., Kajiya, H., Egashira, N., Takekoshi, S., Teramoto, A., et al. (2008) Expression of Wnt4 in human pituitary adenomas regulates activation of the β-catenin-independent pathway. *Endocr. Pathol.*, 19(4): 261–73.
61. Gueorguiev, M. and Grossman, A.B. (2009) Pituitary gland and β-catenin signaling: From ontogeny to oncogenesis. *Pituitary*, 12(3): 245–55.
62. Elston, M.S., Gill, A.J., Conaglen, J.V., Clarkson, A., Shaw, J.M., Law, A.J., et al. (2008) Wnt pathway inhibitors are strongly down-regulated in pituitary tumours. *Endocrinology*, 149(3): 1235–42.
63. Kremenevskaya, N., Khattak, M. and Buchfelder M. (2007) Activation of the Wnt 5A, β-catenin, TCF-4, cyclin D1 and E-cadherin in null cell pituitary adenomas. Poster presentation at the 89th Annual Meeting of the Endocrine Society, June 2–5, 2007, Toronto, Canada, P1-284.
64. Cañibano, C., Rodriguez, N.L., Saez, C., Tovar, S., Garcia-Lavandeira, M., Borrello, M.G., et al. (2007) The dependence receptor Ret induces apoptosis in somatotrophs through a Pit1/p53 pathway, preventing tumor growth. *EMBO J.*, 26(8): 2015–28.
65. Malumbres, M. and Barbacid, M. (2005) Mammalian cyclin-dependent kinases. *Trends Biochem. Sci.*, 30(11): 630–41.
66. Ramsey, M.R., Krishnamurthy, J., Pei, X.H., Torrice, C., Lin, W., Carrasco, D.R., et al. (2007) Expression of p16Ink4a compensates for p18Ink4c loss in cyclin-dependent kinase 4/6-dependent tumours and tissues. *Cancer Res.*, 67(10): 4732–41.
67. Ogino, A., Yoshino, A., Katayama, Y., Watanabe, T., Ota, T., Komine, C., et al. (2005) The p15(INK4b)/p16(INK4a)/RB1 pathway is frequently deregulated in human pituitary adenomas. *J. Neuropathol. Exp. Neurol.*, 64(5): 398–403.
68. Donangelo, I., Marcos, H.P., Araujo, P.B., Marcondes, J., Filho, P.N., Gadelha, M., et al. (2005) Expression of retinoblastoma protein in human growth hormone-secreting pituitary adenomas. *Endocr. Pathol.*, 16(1): 53–62.
69. Vierimaa, O., Georgitsi, M., Lehtonen, R., Vahteristo, P., Kokko, A., Raitila, A., et al. (2006) Pituitary adenoma predisposition caused by germline mutations in the AIP gene. *Science*, 312: 1228–30.
70. Daly, A.F., Vanbellinghen, J.F., Khoo, S.K., Jaffrain-Rea, M.L., Naves, L.A., Guitelman, M.A., et al. (2007) Aryl hydrocarbon receptor-interacting protein gene mutations in familial isolated pituitary adenomas: Analysis in 73 families. *J. Clin. Endocrinol. Metab.*, 92: 1891–96.
71. Toledo, R.A., Lourenco, D.M., Jr., et al., Liberman, B., Cunha-Neto, M.B., Cavalcanti, M.G., Moyses, C.B., et al. (2007) Germline mutation in the aryl hydrocarbon receptor interacting protein gene in familial somatotropinoma. *J. Clin. Endocrinol. Metab.*, 92: 1934–37.
72. Georgitsi, M., De, M.E., Cannavo, S., Makinen, M.J., Tuppurainen, K., Pauletto, P., et al. (2008) Aryl hydrocarbon receptor interacting protein (AIP) gene mutation analysis in children and adolescents with sporadic pituitary adenomas. *Clin. Endocrinol. (Oxf.)*, 69: 621–27.
73. Polyak, K., Li, Y., Zhu, H., Lengauer, C., Willson, J.K., Markowitz, S.D., et al. (1998) Somatic mutations of the mitochondrial genome in human colorectal tumours. *Nat. Genet.*, 20: 291–93.

74. Romero, F., Multon, M.C., Ramos-Morales, F., Dominguez, A., Bernal, J.A., Pintor-Toro, J.A., et al. (2001) Human securin, hPTTG, is associated with Ku heterodimer, the regulatory subunit of the DNA-dependent protein kinase. *Nucleic Acids Res.*, 29: 1300–07.
75. Sotillo, R., Renner, O., Dubus, P., Ruiz-Cabello, J., Martin-Caballero, J., Barbacid, M., et al. (2005) Cooperation between Cdk4 and p27kip1 in tumor development: A preclinical model to evaluate cell cycle inhibitors with therapeutic activity. *Cancer Res.*, 65: 3846–52.
76. Heaney, A.P., Fernando, M. and Melmed, S. (2002) Functional role of estrogen in pituitary tumor pathogenesis. *J. Clin. Invest.*, 109: 277–83.
77. Paez-Pereda, M., Kovalovsky, D., Hopfner, U., Theodoropoulou, M., Pagotto, U., Uhl, E., et al. (2001) Retinoic acid prevents experimental Cushing syndrome. *J. Clin. Invest.*, 108: 1123–31.
78. Hendriks-Stegeman, B.I., Augustijn, K.D., Bakker, B., Holthuizen, P., van der Vliet, P.C. and Jansen, M. (2001) Combined pituitary hormone deficiency caused by compound heterozygosity for two novel mutations in the POU domain of the Pit1/POU1F1 gene. *J. Clin. Endocrinol. Metab.*, 86: 1545–50.
79. Chanock, S.J., Burdett, L., Yeager, M., Llaca, V., Langerod, A., Presswalla, S., et al. (2007) Somatic sequence alterations in twenty-one genes selected by expression profile analysis of breast carcinomas. *Breast Cancer Res.*, 9: R5.
80. Katoh, M. and Katoh, M. (2006) Cross-talk of WNT and FGF signaling pathways at GSK3β to regulate β-catenin and SNAIL signaling cascades. *Cancer Biol. Ther.*, 5: 1059–64.
81. Malumbres, M., Pevarello, P., Barbacid, M. and Bischoff, J.R. (2008) CDK inhibitors in cancer therapy: What is next? *Trends Pharmacol. Sci.*, 29: 16–21.
82. Waizenegger, I., Gimenez-Abian, J.F., Wernic, D. and Peters, J.M. (2002) Regulation of human separase by securin binding and autocleavage. *Curr. Biol.*, 12: 1368–78.
83. Botelho, C.H., Magalhaes, A.V., Mello, P.A., Schmitt, F.C. and Casulari, L.A. (2006) Expression of p53, Ki-67 and c-erb B2 in growth hormone-and/or prolactin-secreting pituitary adenomas. *Arq. Neuropsiquiatr.*, 64: 60–6.
84. Skelly, R.H., Korbonits, M., Grossman, A., Besser, G.M., Monson, J.P., Geddes, J.F., et al. (2000) Expression of the pituitary transcription factor Ptx-1, but not that of the trans-activating factor prop-1, is reduced in human corticotroph adenomas and is associated with decreased α-subunit secretion. *J. Clin. Endocrinol. Metab.*, 85: 2537–42.
85. Pellegata, N.S., Quintanilla-Martinez, L., Siggelkow, H., Samson, E., Bink, K., Hofler, H., et al. (2006) Germ-line mutations in p27Kip1 cause a multiple endocrine neoplasia syndrome in rats and humans. *Proc. Natl. Acad. Sci. U.S.A*, 103: 15558–15563.
86. Garcia-Echeverria, C. and Sellers, W.R. (2008) Drug discovery approaches targeting the PI3K/Akt pathway in cancer. *Oncogene*, 27: 5511–26.
87. Skelly, R.H., Korbonits, M., Grossman, A., Besser, G.M., Monson, J.P., Geddes, J.F., et al. (2000) Expression of the pituitary transcription factor Ptx-1, but not that of the trans-activating factor prop-1, is reduced in human corticotroph adenomas and is associated with decreased α-subunit secretion. *J. Clin. Endocrinol. Metab.*, 85(7): 2537–42.
88. Leng, X., Noble, M., Adams, P.D., Qin, J. and Harper, J.W. (2002) Reversal of growth suppression by p107 via direct phosphorylation by cyclin D1/cyclin-dependent kinase 4. *Mol. Cell Biol.*, 22(7): 2242–54.
89. Garcia-Echeverria, C. and Sellers, W.R. (2008) Drug discovery approaches targeting the PI3K/AKT pathway in cancer. *Oncogene*, 27(41): 5511–26.

7 Stem Cell Markers in Pituitary Adenomas

Features of Stem Cells

Initially considered as being only in the embryos, nowadays there are more data that show the presence of stem cells in adult tissue as well. They are involved in tissue turnover, plasticity, and repair. Therefore, they are present in organs with high regenerative capacity, such as gut, skin, and bone marrow, and also in organs with low turnover and entirely postmitotic organs, such as the brain, heart, retina, and spinal cord [1].

Clonal expansion, self-renewal, differentiation, and preservation of pluripotency are essential characteristics of a stem cell. The capacity of expansion and self-renewal should be demonstrated by at least five passages of the colonies to exclude short-term proliferation activity. Plasticity is important as an adaptative mechanism and also involves stem cells. The pituitary presents an important plasticity during its life span but also in various disorders; this feature certainly involves the stem cells [2].

An important feature of stem cells is the capacity of forming aggregates, called pituispheres, in a similar mode in which the neural stem cells form neurospheres [3]. This capacity was based on the colony-forming assay and was developed from stem cell research [4].

Embryonic and Adult Stem Cells

Stem cells are the source of progenitor cells for tissue, and they present several capacities: self-renewal, proliferation, and differentiation toward adult pituitary cells. Stem cells have long-term self-renewal capacity and asymmetric division, giving rise to one stem cell and one progenitor cell with shorter self-renewal capacity. They represent a cell pool of transit cells that will differentiate toward the other pituitary cell lineages [5]. Considering their spectrum, stem cells can be totipotent, pluripotent, multipotent, oligopotent, or unipotent. Totipotent cells can give rise to a whole embryo, and pluripotent cells can give rise to organs. These are embryonic stem cells, available in the first early days of an organism's life. Multipotent cells can give rise to tissue and oligopotent cells can develop into several cell lineages [4].

Human embryonic stem cells were isolated and described almost 10 years ago. Totipotent and pluripotent cells are obtained from blastomeres by complex microsurgical techniques and cultures, from *in vitro* fertilization procedures [6].

Molecular Pathology of Pituitary Adenomas. DOI: 10.1016/B978-0-12-415830-6.00007-X

Candidates to Stem Cells from Anterior Pituitary: Markers of Stem Cells

In the adult pituitary, multipotent or oligopotent stem cells can be identified in a small proportion (2–5%), being represented by the population of FSCs, follicular cells (FCs), marginal cells (MCs), and cells with mesenchymal phenotype in the so-called side population (SP) [5,7] (Figure 7.1).

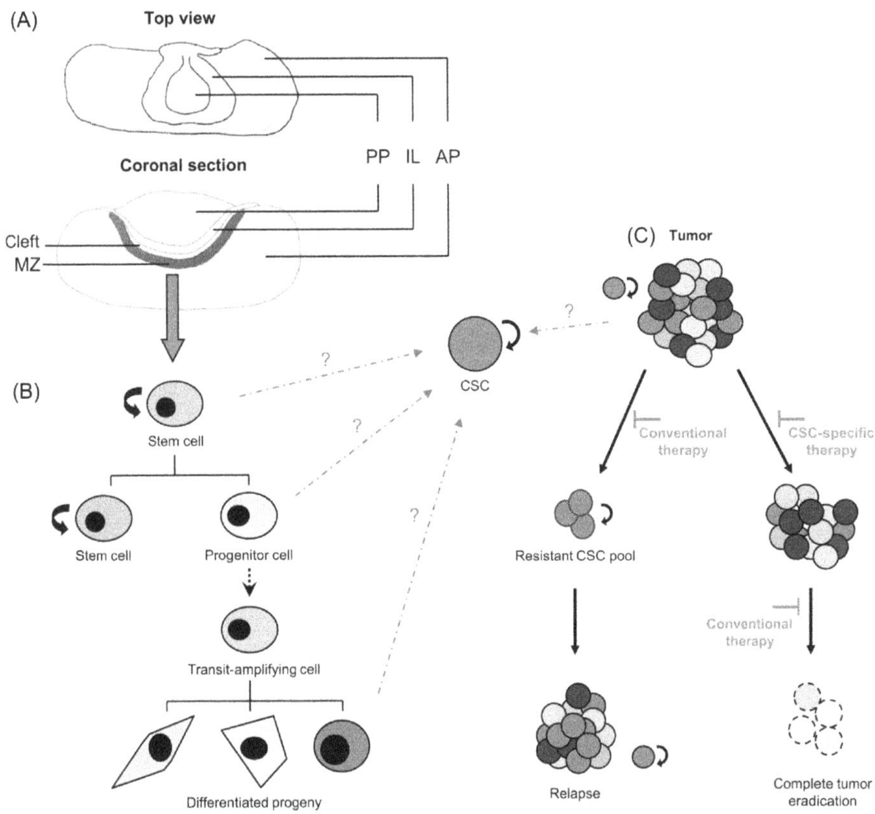

Figure 7.1 Stem/progenitor cells concept in pituitary adenoma. (A) Anterior pituitary (AP), posterior pituitary (PP), intermediate lobe (IL). Periluminal or marginal zone (MZ) hosts the stem/progenitor cells that give rise to the new hormone-producing cells. (B) Stem cells: asymmetric division → one exact-copy stem cell and one progenitor cell with less or no self-renewal capacity. (C) The "cancer stem cell" (CSC) theory: a hierarchical organization of cancer, with a subpopulation of stem cell-like cells at the basis, that drive the initiation, growth, and maintenance of the eventual heterogeneous tumor, and that are also considered responsible for therapeutic resistance as well as metastasis; conventional treatments kill the bulk of the tumor cells but fail to target the CSC, leading to relapse of the cancer. Eradicating the tumor-driving population by means of a CSC-specific therapy may, in combination with conventional therapies, eventually lead to complete destruction of the total tumor mass.

Markers of the stem cells in the pituitary gland are dependent on the group of cells. The FSC markers are angiotensin-converting enzyme (ACE), stem cells antigen 1 (Sca-1), fibronectin, nestin, and SOX-2 (which seems to be the most important) [1]. Follicular cells have no markers that has been demonstrated up to now. MC markers include ACE, Sca-1, nestin, and SOX-2 as cell markers [4]. SP cell markers are ACE, Sca-1, nestin, and CD133. The proportion of SP cells is only 1.5% of anterior pituitary cells, but they are important for pituitary cell turnover. Mesenchymal cell markers are CD133, vimentin, fibronectin, and αSMA [8].

All these cells can generate pituitary cell lineages and represent an important pool for the adult pituitary plasticity. Pituitary adenomas with minimal mitotic activity can be produced by cells that harbor characteristics of stem cells. Animal studies provided data about pituitary stem cells, but their translation into human pituitary tumors is not possible. The secreting capacity is a characteristic of terminal differentiation of the pituitary cells. Due to this feature, stem cells do not exhibit the secreting feature as part of the chromophobe population. The relative importance of these stem cells in generating the different cell lineages is not yet established.

The capacity of a stem cell to generate specific pituitary cell lines was not proved by direct methods *in vitro*, but many indirect arguments do support their role in pituitary plasticity.

Follicullostelate Cells

In the anterior pituitary, in addition to secretory cells, there are cells with an elongated or stellate aspect joined by desmosomes, along with a small nucleus and a small Golgi apparatus showing reduced secretion capacity. They are not secreting hormones, so they are chromophobes. They are supposed to be important for the paracrine regulation of the anterior pituitary (including gonadotroph activity and many other functions), as well as stem cells in the adult gland that are involved in normal cell turnover [9].

FSCs are diffusely distributed in the anterior pituitary and can be detected by S100 and glial fibrillary acid protein (GFAP), in addition to nestin immunostaining, which shares the stem cell ability. Their role as stem cells versus paracrine regulatory cells remains to be elucidated [10]: scavenger activity by phagocytosis of degenerated cells, paracrine regulation by producing various growth factors and cytokines, such as interleukin-6, leukemia inhibitory factor, bFGF, VEGF and follistatin, and large-scale intercellular communication by means of their long cytoplasmic processes and gap junctions.

FSC may have neoplastic potential, as recently discussed by Horvath et al. [9]. The type of pituitary tumor called spindle cell oncocytoma [11] is related to FSC due to immunostaining (vimentin, S100, epithelial membrane antigen (EMA), and galectin3). The clinical aspect is that of nonfunctioning pituitary tumors, but at the ultrastructural level, they resemble fetal pituitary cells at the gestational age of 6–10 weeks. The tumor has oncocytic features and immunophenotype-suggested FSCs as the source cell type. Gonadotroph adenomas, oncocytomas, and null cell adenomas all could originate in FSCs. Nonfunctioning adenomas account for 30–50%

of pituitary adenomas in the elderly [8]. In primary hypothyroidism, TSH cells are obtained by transdifferentiation from somatotrophs, both cell types being obtained from small cells and immunoreactive for either TSH or GH, possibly derived from FSCs. A synthesis of features and pathology data concerning FSC is presented.

Follicular Cells

FCs are cells of irregular shape, with apical microvilli surrounding a small pseudolumina, and the cells are distributed via adenohypophysis [1]. They are distinct from FSCs, despite their similarities. FCs are abundant in embryonic pituitary but scarce in postnatal pituitary. They increase after gonadectomy, suggesting a dependence on the sex steroid feedback. FCs had no specific cell marker, but some authors suggest that they are functional variant of FSC [12].

Marginal Cells

Present between the anterior and intermediate pituitary and closed to the cleft and posterior pituitary, MCs retain undifferentiated characteristics as weak, secretory organelles. They are immunoreactive for nestin and class VI intermediate filament protein, known as a marker of neuronal stem cells but present in other stem/progenitor cells [13]. They express SOX-2, suggesting the stem cell role. However, SOX-2 colocalizes with transcription factors like Pit1 and SF1 and with nestin, which supports the theory that MCs are pluripotent stem cells.

SP Cells

SP cells (SPCs) are related to their capacity of excluding a fluorescent dye, Hoeschst 33342, which binds DNA, a technique developed by Goodell in 1996 for hematopoietic stems [14]. The pituitary SP presents immunoreactivity for nestin and Lhx4, a transcription factor involved in early pituitary development. Lhx4 is important for the survival of progenitor cells in the embryonic pituitary [1]. They express high levels of Sca-1, which are positive for SOX-2 and SOX-9 markers of pituitary progenitor cells [5]. However, a cluster of SPC non-SCA1 displays multipotent differentiation capacity and SOX-2, and generates pituispheres in culture [15]. It seems that SPCs are a heterogeneous population with stem cell–like activity.

From Stem Cells to Pituitary Adenomas

Pituitary adenomas are 10–15% of the intracranial tumors, but few are progressive and most are microadenomas that are associated with normal function and do not do any damage to the surrounding structures. These microadenomas are called *incidentalomas* [16]. The progression from pituitary microadenoma to a big tumor requires the involvement of several mutations as tumor suppressor genes or oncogenes.

On the other hand, invasive tumors tend to release more hormones, in larger amounts, and these features might suggest that the origin of the tumor is an undifferentiated cell.

From this concept to the hypothesis of stem cells as the origin of pituitary tumors, there is a small step. A disruption of the molecular mechanism that controls the normal pituitary stem cell evolution might change the fate of a normal stem cell from normal pituitary cells turnover toward the pituitary tumor.

Oligopotent cells (1–2% of tumors) are important in pituitary adaptation and plasticity. These cells are reservoirs, which can replace the pituitary every 5 weeks, but this never happens in a normal individual. The adult pituitary shows plasticity and adaptation mechanisms in health and disease. The proportion of various subtypes is changing from the neonatal period to puberty and adulthood and during pregnancy and lactation. Various disorders like mixedema or hyperthyroidism, steroid treatment, and primary hypogonadism can change the pituitary percentage of different subpopulations by the so-called transdifferentiation. These changes are related to a stem cell pool and various hormonal factors in the environment from the hypothalamus or target glands. Transcription factors involved in normal pituitary development can be found in pituitary tumors as well, suggesting an influence in pituitary tumorigenesis [17].

The interest in the biology of stem cells is related to their link to pituitary oncogenesis [18]. Nestin+ stem cells may be targeted for pituitary oncogenesis, in addition to the retinoblastoma Rb-1 gene. Crossing nestin mice with Rb+/− mice leads to POMC pituitary tumors in 1-year-old mice. These tumors are surrounded by nestin+ cells that express Lhx3 and SOX-2 but are POMC-negative. This experiment strongly suggests that stem cells can cause pituitary tumors by promoting the development of abnormal cells with high proliferation and secretion capacity [19].

Very recently, Xu et al. [20] reported the isolation of tumor stem-like cells from human pituitary adenomas. For the first time, it is indicated that stem-like cells are present in benign tumors. They obtained pituitary adenoma stem-like cells (PASCs) by cell dissection from human pituitary adenomas and characterized them by self-renewal assays, marker expression analysis, differentiation, and stimulated hormone production assays.

In this study, phenotypical and functional characterization of PASCs from both hormone-positive and -negative pituitary adenomas was performed. It was demonstrated that these PASCs were capable of self-renewal and multipotent differentiation *in vitro*. Both hormone-producing and nonhormone-producing PASCs can be differentiated into hormone-producing cells, with the specific hormones produced determined by the characteristics of the original pituitary tumor. When tumor stem-like cells were injected in Non-Obese Diabetic/Severe Combined Immuno-Deficiency (NOD/SCID) mice, they generated new tumors with the same genetic composition and characteristics as the original tumors. Retransplanted, they maintain the same tumor-specific properties.

Regarding pituitary adenomas, it was demonstrated that PASCs express a range of stem cell–associated markers and two lineage-specific stem/progenitor cell markers: nestin and CD133. Both hormone-producing and nonhormone-producing tumor stem cells forming tumor spheres express these markers. Moreover, sphere-forming pituitary tumor cells express stem/progenitor cell markers, but not differentiated cell markers. The primary culture from tumor tissue was used as a source for all the mentioned assays. In addition, they are characterized in terms of anti-apoptotic proteins and pituitary progenitor cell markers and compared to their differentiated daughter cells. The results of this study shows that PASCs are more resistant to chemotherapeutics and can proliferate *in vivo* in immunotolerant NOD/SCID mice, giving rise to pituitary tumors similar to the primary ones and express a higher level of anti-apoptotic genes. In contrast, differentiated cells are more responsive to hypothalamic hormones and are not able to proliferate *in vivo* toward pituitary adenomas. They conclude that PASCs are self-renewable and multipotent and can initiate transplantable pituitary tumors [20].

The most common mixed pituitary adenoma is the GH–PRL-secreting tumor, mammosomatotroph adenomas being related to the common ancestor of the GH–PRL cell lineage and corresponding transcription factors involved. This is also expressed by the very rare cases of GH–TSH, TSH–PRL, or GH-, PRL-, and TSH-secreting tumors, with the common transcription factor being Pit1.

Another example is related to null cell adenomas, which in fact secrete alpha- or beta-specific glycoprotein subunits that express the SF1 transcription factor. These examples sustained the idea of adenoma progression from uncommitted pituitary stem cells in a certain genetic and hormonal environment. A combination of transcription factors might facilitate the progression or inhibit it. Many null cell adenomas express SF1 and NeuroD1, factors that are present in the pituitary stem cells. The more the transcription factors expressed, the more aggressive the pituitary tumors are. Besides invasive capacity, pituitary tumors might develop metastatic capacity, which is related to the expression of ERs [21].

In a tamoxifen-inducible mouse pituitary tumor model, it was shown that the Pax7+stem cell population can give rise to silent corticotroph adenomas. Pax7 pituitary cells are present in the intermediate lobe. When the Rb gene is deleted, Pax7+cells can transform into ACTH 4 pituitary tumors [22].

Another gene involved in pituitary tumorigenesis and development is the PTTG, encoding a securin protein. Disruption of PTTG can result in chromosomal instability and tumorigenesis. Expression of PTTG in the mouse pituitary can drive multifocal hyperplasia of the mouse pituitary with increase of LH, GH, and TSH, and consequently of IGF1 and testosterone. PTTG overexpression leads to plurihormonal cell focal expansion and pituitary tumors [23].

Concluding Remarks

Stem cells can be involved not only in normal pituitary turnover and adaptation, but also in pituitary adenoma pathogenesis, being a source of a proliferation, if they are subject to oncogenetic mutation. This hypothesis is only the beginning of investigation into this topic and needs more experimental evidence.

References

1. de Almeida, J.P., Sherman, J.H., Salvatori, R. and Quinones-Hinojosa, A. (2010) Pituitary stem cells: Review of the literature and current understanding. *Neurosurgery*, 67(3): 770–80.
2. Vankelecom, H. (2007) Stem cells in the postnatal pituitary? *Neuroendocrinology*, 85(2): 110–30.
3. Nolan, L.A. and Levy, A. (2006) A population of non-luteinising hormone/non-adrenocorticotrophic hormone-positive cells in the male rat anterior pituitary responds mitotically to both gonadectomy and adrenalectomy. *J. Neuroendocrinol.*, 18(9): 655–61.
4. Lepore, D.A., Thomas, G.P., Knight, K.R., Hussey, A.J., Callahan, T., Wagner, J., Morrison, W.A. & Thomas, P.Q. (2007) Survival and differentiation of pituitary colony-forming cells *in vivo*. *Stem Cells*, 25(7): 1730–36.
5. Vankelecom, H. and Gremeaux, L. (2010) Stem cells in the pituitary gland: A burgeoning field. *Gen. Comp. Endocrinol.*, 166(3): 478–88.
6. Sills, E.S., Takeuchi, T., Tanaka, N., Neri, Q.V. and Palermo, G.D. (2005) Identification and isolation of embryonic stem cells in reproductive endocrinology: Theoretical protocols for conservation of human embryos derived from *in vitro* fertilization. *Theor. Biol. Med. Model*, 2: 25.
7. Vankelecom, H. (2007) Non-hormonal cell types in the pituitary candidating for stem cell. *Semin. Cell Dev. Biol.*, 18(4): 559–70.
8. Horvath, E. and Kovacs, K. (2002) Folliculostellate cells of the human pituitary: A type of adult stem cell? *Ultrastruct. Pathol.*, 26(4): 219–28.
9. Horvath, E., Coire, C.I., Kovacs, K. and Smyth, H.S. (2010) Folliculo-stellate cells of the human pituitary as adult stem cells: Examples of their neoplastic potential. *Ultrastruct. Pathol.*, 34(3): 133–39.
10. Devnath, S. and Inoue, K. (2008) An insight to pituitary folliculo-stellate cells. *J. Neuroendocrinol.*, 20(6): 687–91.
11. Roncaroli, F., Scheithauer, B.W., Cenacchi, G., Horvath, E., Kovacs, K., Lloyd, R.V., bell-Aleff, P., Santi, M. & Yates, A.J. (2002) “Spindle cell oncocytoma” of the adenohypophysis: A tumor of folliculostellate cells? *Am. J. Surg. Pathol.*, 26(8): 1048–55.
12. Allaerts, W. and Vankelecom, H. (2005) History and perspectives of pituitary folliculo-stellate cell research. *Eur. J. Endocrinol.*, 153(1): 1–12.
13. Krylyshkina, O., Chen, J., Mebis, L., Denef, C. and Vankelecom, H. (2005) Nestin-immunoreactive cells in rat pituitary are neither hormonal nor typical folliculo-stellate cells. *Endocrinology*, 146(5): 2376–87.
14. Goodell, M.A., Brose, K., Paradis, G., Conner, A.S. and Mulligan, R.C. (1996) Isolation and functional properties of murine hematopoietic stem cells that are replicating *in vivo*. *J. Exp. Med.*, 183(4): 1797–806.
15. Chen, J., Gremeaux, L., Fu, Q., Liekens, D., Van Laere, S. and Vankelecom, H. (2009) Pituitary progenitor cells tracked down by side population dissection. *Stem Cells*, 27(5): 1182–95.

16. Saeger, W. (1996) Current pathological classification of pituitary adenomas. *Acta. Neurochir. Suppl.*, 65: 1–3.
17. Levy, A. (2008) Stem cells, hormones and pituitary adenomas. *J. Neuroendocrinol.*, 20(1): 139–40.
18. Morrison, S.J. and Kimble, J. (2006) Asymmetric and symmetric stem-cell divisions in development and cancer. *Nature*, 441(7097): 1068–74.
19. Gleiberman, A.S., Michurina, T., Encinas, J.M., Roig, J.L., Krasnov, P., Balordi, F., Fishell, G., Rosenfeld, M.G. & Enikolopov, G. (2008) Genetic approaches identify adult pituitary stem cells. *Proc. Natl. Acad. Sci. U.S.A.*, 105(17): 6332–37.
20. Xu, Q., Yuan, X., Tunici, P., Liu, G., Fan, X., Xu, M., Hu, J., Hwang, J. Y., Farkas, D. L., Black, K.L. & Yu, J.S. (2009) Isolation of tumour stem-like cells from benign tumours. *Br. J. Cancer*, 101(2): 303–11.
21. Ishii, Y., Suzuki, M., Takekoshi, S., Yamazaki, M., Miyai, S., Sanno, N., Teramoto, A. & Osamura, R.Y. (2006) Immunonegative "null cell" adenomas and gonadotropin (Gn) subunit (SUs) immunopositive adenomas share frequent expression of multiple transcription factors. *Endocr. Pathol.*, 17(1): 35–43.
22. Hosoyama, T., Nishijo, K., Garcia, M.M., Schaffer, B.S., Ohshima-Hosoyama, S., Prajapati, S.I., Davis, M.D., Grant, W.F., Scheithauer, B.W., Marks, D.L., Rubin, B.P. & Keller, C. (2010) A postnatal Pax7 progenitor gives rise to pituitary adenomas. *Genes Cancer*, 1(4): 388–402.
23. Abbud, R.A., Takumi, I., Barker, E.M., Ren, S.G., Chen, D.Y., Wawrowsky, K. & Melmed, S (2005) Early multipotential pituitary focal hyperplasia in the alpha-subunit of glycoprotein hormone-driven pituitary tumor-transforming gene transgenic mice. *Mol. Endocrinol.*, 19(5): 1383–91.

8 MicroRNAs

MicroRNAs (miRNAs) represent a class of small, 18- to 28-nucleotide-long, noncoding RNA molecules. Until now, 940 members of the family were identified in humans. Their major role is in the posttranscriptional regulation of protein expression, and their involvement was demonstrated in normal and in pathological cellular processes. miRNAs can be described as "multivalent," with one miRNA able to target multiple genes, thus regulating the expression of several proteins. They were demonstrated to act on several key cellular processes, such as cell differentiation, cell cycle progression, and apoptosis. In tumors, some miRNAs function as oncogenes, others as tumor suppressors; upregulation of oncogenic miRNAs (oncomiRs) was demonstrated in cancer cells (Figure 8.1).

In cancer, miRNAs were found to "overlap" the oncogenesis entirely; some of them were found "upstream," within the triggers of carcinogenesis or progression, while others appear "downstream," as outcomes of oncogenic transformation and/or progression. They were demonstrated to have relevant roles in the multistep processes of oncogenesis, expressing either oncogenic or tumor-suppressor functions.

Recently, it has been shown that several human cancers (e.g., breast, digestive tract, lung, brain, thyroid, and hematologic malignancies) are associated with altered miRNA expression [1]. MiRNAs were demonstrated to play relevant roles in pituitary differentiation.

Currently, many researchers are preoccupied by the miRNA role in pituitary adenomas. The review by Zatelli et al. [2] reports the current knowledge of miRNA expression in pituitary adenomas, focusing on recent microarray data. Moreover, they provided a discussion concerning the possible role of validated and putative targets of the most dysregulated miRNA in pituitary adenoma pathogenesis [2].

Using the microarray technique, several deregulated miRNAs have been involved not only in pituitary cell proliferation and apoptosis but also in neoplastic transformation [3]. Bottoni et al. [4] found that miR-15a and miR-16-1 are expressed at lower levels in pituitary adenomas than in normal pituitary tissue. The miRNAs are small, noncoding RNAs, functioning as antisense regulators of other RNAs. MiR-15a and miR-16-1 genes are located at chromosome 13q14, a region that is frequently deleted in pituitary tumors [4]. Downregulation of these miRNAs in pituitary adenomas correlates with a greater tumor diameter and a lower p43 secretion (as cofactors influence the activity of arginyl-tRNA-synthetase, which is involved in inflammation and angiogenesis), suggesting that these genes may influence tumor growth, at least in part.

These data led to the hypothesis that miR-15a and miR-16-1 function as tumor suppressors and that their inactivation by allelic loss may contribute to tumorigenesis.

Bottoni et al. [4] demonstrated an inverse correlation between tumor diameter and the expression level of miR-15a and miR-16 in samples of GH- or PRL-secreting

Molecular Pathology of Pituitary Adenomas. DOI: 10.1016/B978-0-12-415830-6.00008-1

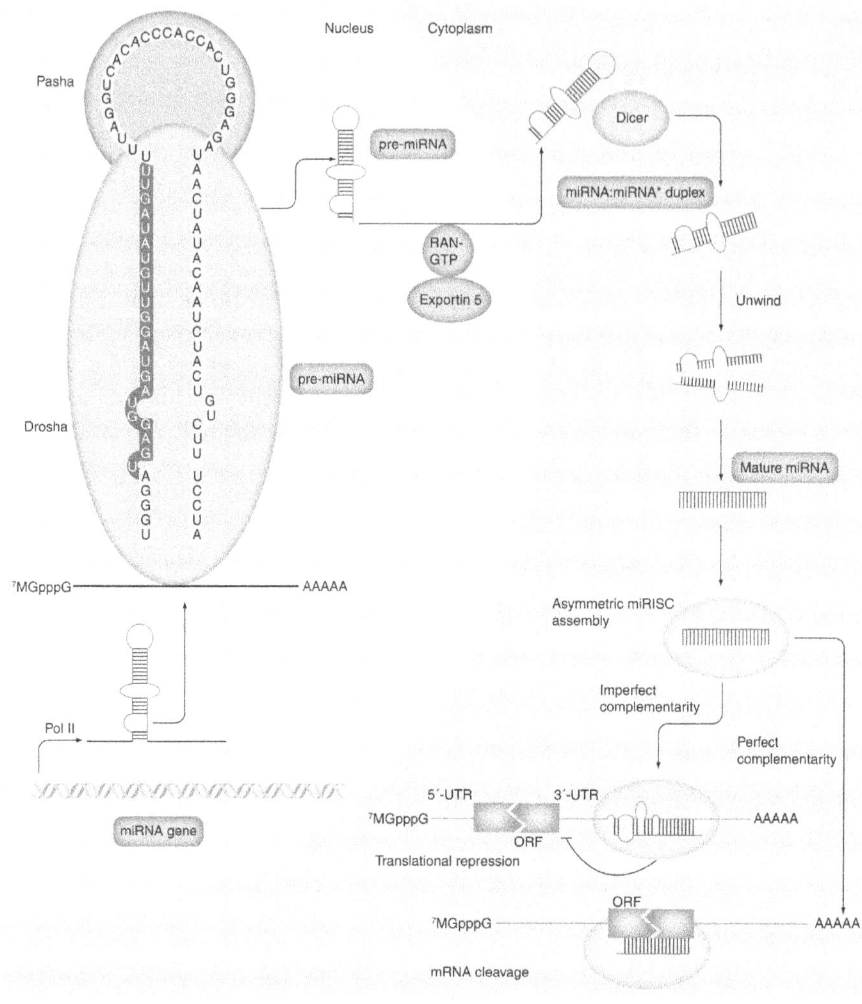

Figure 8.1 Representation of miRNA biogenesis and action. UTR, untranslated region; ORF, open reading frame; PolII, polymerase II; RISC, RNA-induced silencing complex. *Source*: Redrawn with permission from [13].

pituitary adenomas. In short, these findings suggested a role of reduced expression of miR-15a and miR-16 in the pathogenesis of pituitary tumors.

No correlation between loss of the RB (retinoblastoma protein) gene, which is located in the vicinity, and deregulation of miR-15a and miR-16-1 has been reported, indicating that the loss of miR-51a and miR-16-1 expression is often independent of the absence of the gene encoding RB [5]. A study by Amaral et al. [6] analyzed the differential expression of let-7a, miR-15a, miR-16, miR-21, miR-141, miR-143, miR-145, and miR-150 in corticotropinomas and normal pituitary tissue and verified whether their profile of expression correlated with tumor size or remission after treatment. They found underexpression of miR-145, miR-21, miR-141, let-7a, miR-150, miR-15a, miR-16, and miR-143 in ACTH-secreting pituitary tumors when compared to normal pituitary tissues. There were no differences between miRNA expression and tumor size or miRNA expression and ratio of remission after surgery, except in patients presenting lower miR-141 expression, who showed a better chance of remission. This suggests a possible role of miR-141 in the regulation of pituitary genes involved in tumor growth and tumor local invasion. The results that altered the expression of miRNAs are involved in GH-secreting pituitary adenoma transformation, which will shed light on the mechanisms for the treatment of acromegaly by somatostatin analogs (SSAs). Identification and characterization of the targets of altered miRNA genes may elucidate molecular mechanisms involved in the pathogenesis of pituitary adenomas [6].

The results of Amaral et al. support the possibility that altered miRNA expression profile might be involved in corticotrophic tumorigenesis. However, the lack of knowledge about miRNA target genes postpones full understanding of the biological functions of downregulated or upregulated miRNAs in corticotropinomas.

Other studies have demonstrated that several identified miRNAs are involved in cell proliferation, apoptosis, and corticotrophic tumorigenesis, suggesting that deregulation of miRNA expression may be involved in pituitary tumorigenesis [6,7]. Decreased expression of let-7 was confirmed in 23 of 55 adenomas (42%) and was correlated with high-grade tumors. An inverse correlation between let-7 and high-mobility group A2 expression was evident. These findings support a causal link between let-7 and high-mobility group A2, whereby loss of let-7 expression induces high-mobility group A2 upregulation that represents an important mechanism in pituitary tumorigenesis and progression [8].

Mao et al. examined miRNA expressions in several pituitary adenomas. A total of 52 miRNAs are differentially expressed between GH-secreting adenomas and non-GH-secreting ones. Of which, 23 overexpressed miRNAs (with levels of overexpression ranging from 2 to 8) were found, while 29 miRNAs were downregulated, with levels of expression between 0.04 and 0.5.

The highest overexpressions were found for miR-365 and 768-3p. They have also reported a total of nine miRNAs differentially expressed between macroadenomas and microadenomas. Cluster analysis based on these differentially expressed miRNAs showed that macroadenomas and microadenomas belonged to two distinct groups. Among nine differentially expressed miRNAs between micro- and macro-GH-secreting pituitary adenomas, the expression of miR-15a was downregulated. A total of 13 miRNAs were differentially expressed between the GH-secreting pituitary adenomas with lanreotide-treated patients and those without lanreotide treatment. They also found that seven miRNAs were differentially expressed between SSA responders and SSA nonresponders. Cluster analysis based on these differentially expressed miRNAs showed a clear distinction between SSA responders and SSA nonresponders.

miR-126 and miR-381 were downregulated in GH-secreting pituitary adenomas compared to normal pituitary. Loss of miR-126 (targeting p85 subunit of IP3K [9]) results in amplification of the IP3 signal, facilitating tumorigenesis. The target of miR-126 and miR-381 is PTTG protein 1, which is involved in multiple cellular pathways, including proliferation, DNA repair, transformation, angiogenesis induction, invasion, and the induction of genetic instability. PTTG is overexpressed in most pituitary adenomas and is correlated to the recurrence and angiogenesis [10]. Their results, therefore, indicated that altered expression of miR-126 genes may play an important role in the development of GH-secreting pituitary adenomas.

However, miR-524-5p was downregulated in SSA responders compared to non-SSA responders. In contrast, miR-524-5p was upregulated in the lanreotide-treated patients compared to the untreated patients. The possible reason of miR-524-5p up- and downregulation in different groups is due to the different responsiveness to SSA treatment in various cases.

Thus, in GH-responder cases, the level of growth factor is high. Lanreotide is a type of inhibitory growth factor. It is possible that miR-524-5p is negatively correlated with growth factors; e.g., lanreotide promotes miR-524-5p upregulation and inhibits the growth factors. The similar reason may explain the opposite phenomenon occurred with the miR-193a-5p, miR-574-5p, miR-96, and miR-99b. However, further studies are needed to elucidate the function and mechanisms of altered expression of miR-524-5p.

These results also implied that establishing a novel, evolutionarily conserved strategy to keep the balance between miRNAs and their transcriptional regulatory programs is necessary. miR-145 was downregulated in GH-secreting pituitary adenomas compared to normal pituitary. Amaral et al. [6] observed that miR-145 was downregulated in 11 samples of corticotropinomas, suggesting a possible role of miR-145 in carcinogenesis. The potential target genes of miR-145 encode oncogenic proteins, such as myc, kras, fos, yes, fli, cyclin D2, and MAPK transduction proteins. miR-145 targets the insulin receptor substrate-1 (IRS-1) and miR-151-3p targets the IRS-4, which regulated cell communication, receptor, and membrane activities.

Some of the miRNAs in the study of Mao et al., such as miR-769-5p, miR-885-5p, miR-886-5p, and miR-890, are newly discovered in pituitary adenoma samples because the new array contains more capture probes and their functions are unknown. The differentially expressed miRNAs are correlated with adenoma characteristics.

Based on the results regarding differentially expressed miRNAs between GH-secreting pituitary adenomas and normal pituitary samples, Mao et al. [11] emphasizes the potential involvement in cell growth, apoptosis, cell proliferation, and tumor development. Furthermore, some differentially expressed miRNAs are associated with tumor diameter, lanreotide treatment, and responsiveness to SSA. Studying the targets of deregulated miRNAs may elucidate molecular mechanisms involved in pituitary adenoma pathogenesis.

Lately, there were more miRNAs up- (188) or downregulated (160) between adenomas and normal pituitaries compared to carcinomas and normal pituitaries (92 up- and 91 downregulated) or between carcinomas and adenomas (46 up- and 52 downregulated). Both RT-PCR (real-time polymerase chain reaction) and *in situ* hybridization showed significant upregulation of miRNA-122 between pituitary carcinomas and adenomas. miRNA-493 was also upregulated in carcinomas compared to ACTH adenomas. Analysis of genes that miRNA-493 interacts with included LGALS3 (lectin, galactose binding, soluble 3) and RUNX2 (runt-related transcription factor 2) both of which have been shown to have roles in pituitary tumor cell growth. These results provide information about marker miRNAs that may lead to further insight into the regulation of pituitary tumor growth and development [12].

In conclusion, the results obtained from pituitary adenomas support the possibility that altered miRNA expression profiles might be involved in pituitary tumorigenesis. However, the lack of knowledge about miRNA target genes postpones full understanding of the biological functions of miRNAs. Therefore, further studies are needed to predict miRNA target genes for either down- or upregulated miRNAs in pituitary adenomas. Predictive miRNAs could be potentially useful diagnostic markers, improving the classification of pituitary adenomas.

References

1. Albulescu, R., Neagu, M., Albulescu, L. and Tanase, C. (2011) Tissular and soluble miRNAs for diagnostic and therapy improvement in digestive tract cancers. *Expert Rev. Mol. Diagn.*, 11(1): 101–20.
2. Zatelli, M.C. and degli Uberti, E.C. (2008) MicroRNAs and possible role in pituitary adenomas. *Semin. Reprod. Med.*, 26(6): 453–60.
3. Bottoni, A., Zatelli, M.C., Ferracin, M., Tagliati, F., Piccin, D., Vignali, C., Calin, G.A., Negrini, M., Croce, C.M., and Degli Uberti, E.C. (2007) Identification of differentially expressed microRNAs by microarray: A possible role for microRNA genes in pituitary adenomas. *J. Cell Physiol.*, 210: 370–77.
4. Bottoni, A., Piccin, D., Tagliati, F., Luchin, A., Zatelli, M.C. and degli Uberti, E.C. (2005) miR-15a and miR-16-1 downregulation in pituitary adenomas. *J. Cell Physiol.*, 204: 280–85.
5. Aqeilan, R.I., Calin, G.A. and Croce, C.M. (2010) XmiR-15a and miR-16-1 in cancer: Discovery, function and future perspectives. *Cell Death Differ.*, 17(2): 215–20.
6. Amaral, F.C., Torres, N., Saggioro, F., Neder, L., Machado, H.R., Silva, W.A., Moreira, A.C., and Castro, M. (2009) MicroRNAs differentially expressed in ACTH-secreting pituitary tumors. *J. Clin. Endocrinol. Metab.*, 94: 320–23.
7. Sivapragasam, M., Rotondo, F., Lloyd, R.V., Scheithauer, B.W., Cusimano, M., Syro, L.V., Kovacs, K., (2011) MicroRNAs in the Human Pituitary. *Endocr. Pathol.*, 22(3): 134–43.
8. Qian, Z.R., Asa, S.L., Siomi, H., Siomi, M.C., Yoshimoto, K., Yamada, S., Wang, E.L., Rahman, M.M., Inoue, H., Itakura, M., Kudo, E., and Sano, T. (2009) Overexpression of HMGA2 relates to reduction of the let-7 and its relationship to clinicopathological features in pituitary adenomas. *Mod. Pathol.*, 22(3): 431–41.
9. Guo, C.G., Sah, J.F., Beard, L., Willson, J.K.V., Markowitz, S.D. and Guda, K. (2008) The noncoding RNA, miR-126, suppresses the growth of neoplastic cells by targeting phosphatidylinositol 3-kinase signaling and is frequently lost in colon cancers. *Genes Chromosomes Cancer*, 47: 939–46.
10. Salehi, F., Kovacs, K., Scheithauer, B.W., Lloyd, R.V. and Cusimano, M. (2008) Pituitary tumor transforming gene in endocrine and other neoplasms: A review and update. *Endocr. Relat. Cancer*, 15: 721–43.
11. Mao, Z.G., He, D.S., Zhou, J., Yao, B., Xiao, W.W., Chen, C.H., Zhu, Y.H., and Wang H.J. (2010) Differential expression of microRNAs in GH-secreting pituitary adenomas. *Diagn. Pathol.*, 5: 79.

12. Stilling, G., Sun, Z., Zhang, S., Jin, L., Righi, A., Kovacs, G., Korbonitis, M., Scheithauer, B.W., Kovacs, K., and Lloyd, R.V. (2010) MicroRNA expression in ACTH-producing pituitary tumors: Up-regulation of microRNA-122 and -493 in pituitary carcinomas. *Endocrine*, 38(1): 67–75.
13. Esquela-Kerscher, A., Slack, F.J. (2006) Oncomirs. microRNAs with a role in cancer. *Nat. Rev. Cancer*, 6(4): 259–269.

9 New Generic Tools for Diagnosis: Genomics and Proteomics

Numerous new techniques are being rapidly developed and applied in the field of normal and neoplastic pituitary.

The cancer biomarkers have been defined as a substance found in an altered amount in the body, its presence being associated to different types of neoplasia. Unfortunately, the development of tumor-associated serum protein biomarkers over the past few decades has not been effective for diagnosing primary neoplasia. There is a considerable pressure to discover new disease-related biomarkers. Identification of diagnostic biomarkers will be essential for improved early intervention in disease and will be a key technology in the development of more-focused drug prescribing.

Biomarkers are also quantitative measures of biological effects that provide informative links between the mechanism of action and clinical effectiveness. They can provide new insights into a drug's mechanism of action, metabolism, efficacy, and safety, and into disease mechanisms and disease course [1].

In order for this vision to be achieved, the right approach to optimization of biomarker investment, performance, and application is needed. This is a deliverable core of translational medicine research [2]. Multiple investigative instruments have been "launched" during biomarker discovery. Emerging omic technologies are among the best suited to approach the domain.

The long-term goals for this human pituitary study are to clarify the molecular mechanisms that are involved in pituitary adenoma pathogenesis and to discover tumor biomarkers. Knowledge of significant signaling pathway networks will provide important clues and clear directions for an in-depth investigation of pituitary adenomas, for the discovery of tumor biomarkers, and for the development of efficacious therapeutic agents.

Over the past several decades, high-throughput "omic" technologies (including genomics, transcriptomics, and proteomics) have been used in many fields, including biology and human diseases. Relative to the traditional molecular biology methods that had been used to study the role of a single gene, single protein, or single small-molecule model, those "omic" data have driven the rapid development of systems biology to study a multiple-factor model of disease and to address the network of interaction and regulatory events that contribute to a disease. Pathway biology, as one important component of systems biology, has been extensively developed. Omic data-based pathway biology relies on an accurate and effective pathway analysis system. Proteomic data obtained from pituitary adenoma tissues were analyzed with Ingenuity Pathway Analysis (IPA) software to define which well-characterized cell-signaling and metabolic pathways could be the most relevant during pituitary adenoma pathogenesis [3].

Molecular Pathology of Pituitary Adenomas. DOI: 10.1016/B978-0-12-415830-6.00009-3

Genomic analysis is one of the most intensively exploited "instrument" in biomarker discovery; it is attacking the discovery of biomarkers by a large array of distinct, "analyte-oriented" approaches, which include DNA sequencing, RNA expression profiling, epigenetics, and miRNA [4]. *DNA sequencing* is frequently used to identify changes in candidate genes. The new generation of sequencing technologies has accelerated the discovery process of DNA sequence variants with clinical significance. The detection of DNA sequence versions, or small insertions or deletions in genes, is now entering clinical practice. Oncogenic DNA sequence variants became diagnostic biomarkers that provide specific therapeutic targets. Large-scale sequencing of candidate genes provides new biomarkers for tumor diagnosis and prognosis. Somatic DNA sequence analysis has recently expanded to genome-wide high-performance DNA sequencing leading to whole genome analysis of DNA sequence variations in tumors [5–7].

The *mRNA expression* is used to study gene expression profiles as relevant biomarkers for diagnosis and prognosis of tumors. Gene expression profiles containing panels of mRNAs can be more effective than standard methods for patient stratification, such as histopathology biomarkers. Biomarker discovery usually starts with large-scale gene profiling (on microarrays) with a technological progression for clinical application to simple microarrays with fewer samples or multiplexed quantitative RT-PCR. There are numerous software approaches to analyzing pathways for a given gene expression dataset. Those that weigh the impact of the particular genes provide additional guidance to the identification of shorter lists of biomarkers [8,9].

The *epigenetic modifications* are the most recent biomarker candidates, where groups of CpG methylation events in promoters are easily detected by PCR-sequencing of DNA from cells or tumors, or by methylation-specific PCR techniques. DNA sequencing is providing whole genome approaches to discover novel epigenetic changes at the DNA methylation level. Aberrant promoter methylation has been found in growth-suppressive genes in human tumorigenesis [10,11].

Due to their small size, miRNAs have an elevated potential as biomarkers because they are easily quantified in normal and tumoral tissues, as well as body fluids. However, this field is still in a very early stage; thus, no miRNA biomarkers are in clinical use [12,13].

Proteomics

Protein biomarkers show a high potential of conversion into paraclinical diagnostic tests. They are often identified in basic science studies of neoplastic cells as overexpressed proteins. Tumor-specific alterations in proteins may occur at the level of protein abundance or posttranslational protein modification, such as glycosylation or phosphorylation [4]. There is a number of direct approaches to identify specific changes in proteins, including abundance or posttranslational modifications. Because proteomics is one of the "focal areas" in scientific, bioinformatics, and technological development, rapid progress was recorded, and a broad spectrum of laboratory technologies have been designed and put into operation. In the post-genome era, some overwhelming issues, such as sequencing or precise identification of posttranslational modification sites, became widely available.

Proteomic changes in neoplasia can be identified by *2D gel electrophoresis* (2DGE). In this technique, different visualization approaches may be applied, including radioactive labeling, covalent attachment of fluorescent tags, and silver staining. Significant enhancements seem to be provided by the recent development of DIGE (difference gel electrophoresis), which combines the ease and sensitivity of detection with the simultaneous processing of samples and controls. 2D and 2D-DIGE are used either separately or in combination with other platforms (such as liquid chromatography/mass spectrometry (LC/MS) or matrix-assisted laser desorption and ionization (MALDI)) to elucidate and identify relevant proteins. However, this technique does not seem to provide the maximum output in the discovery stage, and even advanced instruments have a low sample capacity when compared to other discovery tools. Still, 2D and even 2D-DIGE to a greater degree represent inestimable (and almost irreplaceable) tools in the structural identification/sequencing step, just by facilitating the next step: the development of molecular recognition diagnostics tools.

Mass spectrometric analysis can represent a stand-alone approach for biomarker discovery, or a second stage, for post-electrophoretic sequence identification. 2D-MALDI platforms already have a long history in biomarker discovery [13], and they still represent a major player.

Proteomic MS–based platforms have provided the ability to identify large numbers of novel proteins that have the potential to be biomarkers. These studies are being performed in various biological matrices, such as cultures of normal and tumor cells or human clinical samples (serum, plasma, and urine). Novel biomarker discovery technologies are developed using quantitative isotopic labeling, improved mass spectroscopy algorithms, and improved sample preparation [14]. Isotopic labeling is exploited for relative quantitation of proteins by MALDI with time-of-flight (MALDI-TOF) or LC-MS/MS (mass-spectrometry) high-resolution MS [15,16].

New methods for rapid identification of both known and unknown proteins are poorly developed. MALDI-TOF MS and surface-enhanced laser desorption and ionization with TOF spectrometry (SELDI-TOF) are two of the methods being currently used. MALDI techniques immobilize protein/peptide samples in an energy-absorbing chemical matrix on a chip or plate. MALDI analysis is well suited for the resolution of proteins m/z values 20.000, covering mostly the low-molecular-weight proteome, a hitherto poorly dissected information reserve. It proves relevant resolving power for peptide fragments, in association with bioinformatic tools, while recent systems are also able to perform *de novo* sequencing.

Conversely, SELDI technology uses selective surfaces for binding a subset of proteins based on absorption, partition, electrostatic interaction, or affinity chromatography on a solid-phase protein chip surface [17,18]. The TOF is affected by the mass of the particle and the charge it bears (i.e., the m/z ratio). The detector plate records the intensity of the signal at a given m/z value, and a spectrum is generated. The different peaks in the spectrum correspond to different m/z protein species. This data stream of information can be coupled with data streams from a series of test subjects and complex bioinformatics to define discriminants for neoplasia detection.

With focus on the discovery stage, TOF MS represents a mature and solid platform; both MALDI and SELDI (even if criticized for their less-sensitive detectors) are convenient because they prove to be robust labor- and time-saving solutions, suitable for high-throughput application. They appear to be an almost perfect match for the biomarker discovery and validation strategies, where group sizes range from tens to several hundreds of subjects.

Electrospray Ionization-MS/MS

Electrospray ionization tandem MS (ESI-MS) uses a more discrete ionization compared to other MS techniques; therefore, the protein fragmentation is less extended. For structural studies, the molecule is further split and analyzed in the tandem technology (namely, ESI-MS coupled with MS [19]). Recently, using this technology in the sera of patients diagnosed with pancreatic cancer, various candidate proteins have been discovered, such as α-2 macroglobulin, ceruloplasmin, and C3 component of the complement system [20]. ESI-MS/MS recently evolved to the so-called nanospray ionization, while "tandem" MS is rapidly replaced by "cascades" of mass spectrometers using more sophisticated and sensitive detectors (such as orbit traps), and accommodating other intermediate "sequencing" instruments, like electron-transfer dissociation.

Although the "technological" progress is outstanding, some of the inherent features of most proteomes make them too complex to be dealt with by a single instrumental approach. Most often, a multiplatform approach is needed for the "elucidation" step in order to reduce sample complexity and dynamic range.

Microarrays

Multianalyte protein detection in complex matrices is approached using highly parallel immunoassays on antibody microarrays that resemble highly parallel ELISAs (enzyme-linked immunosorbent assay). There are a few approaches for this type of biomarker discovery tool [21,22].

Solid and fluid microarrays, having a high multiplexing potential, represent a convenient solution in the simultaneous detection of multiple analytes. Because detection is based on epitope recognition, the set of available analytes is limited to that of existing suitable antibodies. It is also limited by some other issues, such as cross-reactivity. However, once a certain set of biomarkers is discovered and elucidated by the use of other platforms, specific microarrays can be developed, providing a straightforward diagnostic solution.

Genomics in Pituitary Adenomas

DNA microarrays represent an efficient method for simultaneously assessing the expression levels of thousands of genes, identifying disease subphenotypes, and predicting pituitary tumor progression [23]. There are few studies that have applied microarray technology to the field of human pituitary tumors. It is expected, though, that microarray technology applications could have relevant results in human pituitary adenoma compared to studies in other solid tumors. The potential of this gene analysis, or more accurately "transcriptome," could lead wide parallel analysis toward identification of genes that underlie tumorigenesis, which are prognostic markers or therapeutic targets [24].

Jiang et al. [25] have used a fiber-optic BeadArray to analyze gene expression in prolactinomas. Their results showed 27 upregulated and 182 downregulated genes, of which they chose 4 for validation by RT-PCR. Further pathway analysis has shown implication of the P53- and GnRH-signaling pathways in the tumorigenesis of prolactinomas. Their data suggest that fiber-optic BeadArray, combined with pathway analysis of differential gene expression profiles, is a valid approach for investigating the pathogenesis of tumors [25].

Anterior studies of Evans et al. [26] and Moreno et al. [27] identified multiple transcripts as being significantly overexpressed or underexpressed in tumors relative to normal pituitary and validated selected transcripts by quantitative RT-PCR. Evans et al. used cDNA arrays in secretory and nonfunctioning human pituitary adenomas to identify gene expression profiles. Their studies revealed differential expression of 188 genes out of 7075. Of these genes, 60 were differentially expressed in NF adenomas, 47 genes in PRL-secreting adenomas, 30 genes in GH-secreting adenomas, and 51 genes in ACTH-secreting adenomas, respectively. Three of these genes were chosen for validation by RT-PCR.

Galland et al. [28] performed pangenomic analysis using an oligonucleotide microarray in order to discriminate between gene expression in invasive and noninvasive NFPA. It was shown that 346 genes differed between invasive and noninvasive NFPAs, out of which 233 genes were upregulated and 113 genes were downregulated in invasive tumors. In addition, 35 genes were selected for expression quantification by qRT-PCR, but only 4 genes were confirmed as overexpressed: IGFBP5, MYO5A, FLT3, and NFE2L1. Out of these, only myosin 5A (MYO5A) immunostaining was stronger in invasive than in noninvasive tumors, implying that it is potentially a useful marker of tumor invasiveness [28].

Proteomics and Transcriptomics

Proteomic methods included 2DGE, 2D gel image analysis, MS (MALDI-TOF peptide mass fingerprinting (MALDI-TOF PMF) and LC-electrospray ionization-quadruple-ion trap tandem MS (LC-ESI-Q-IT MS/MS)), and database analysis. Transcriptomics methods included the GeneChip microarray, image processing, and data analysis.

The comparative-proteomics strategy, which combines 2DGE, digitized 2D gel image analysis, MS, and bioinformatics analysis, has been used to locate and characterize differentially expressed proteins (DEPs) between human pituitary adenoma tissue and controls. The high level of reproducibility and a wide dynamic separation range of a 2DGE system were obtained for comparative proteomics [29,30].

The first study regarding the applications of proteomics and the realizations of protein profiling has been realized by Beranova-Giorgianni et al. [31] in the human pituitary. Among the identified proteins, there are pituitary hormones, structural proteins, and enzymes [31].

In order to clarify the basic molecular mechanisms that participate in the formation of human pituitary macroadenomas, Desiderio et al. [32] described the comparative proteomics between a pituitary adenoma tissue and a control tissue for the first time. A vertical, 2D polyacrylamide gel electrophoresis system and PDQuest image analysis software were used to provide a high level of between-gel reproducibility and electrophoretic separation to locate each DEP accurately. MS (MALDI-TOF and LC-ESI-Q-IT) and protein databases were used to characterize each DEP. A total of 137 differential gel spots (37 increased spot volumes, 39 decreased, 19 new, and 42 lost) were found when we compared an adenoma proteome to a control proteome. A total of 71 spots (20 increased, 27 decreased, 13 new, and 11 lost), representing 39 differentially regulated proteins, were identified. *Five differentially regulated proteins (PRL, cellular retinoic acid–binding protein II, G-protein β-subunit 3, secretagogin, and calreticulin) were also validated with results from a comparative transcriptomics study of pituitary adenomas and controls.* The functional characteristics of these DEPs provide a differential proteomic profile between a pituitary adenoma and a control [32].

In order to explore the potential role of secretagogin in pituitary adenomas, an analytical strategy has been used, strategy that combines proteomic and transcriptomics. The proteomics and transcriptomics data demonstrated that secretagogin was significantly downregulated at the protein and mRNA levels, respectively, in the human NFPAs (NF−, LH+, FSH+, and FSH++ LH+). The secretagogin protein expression correlated significantly with its mRNA expression. Those results suggest that secretagogin might play a role in human NFPAs. This novel finding may provide clues to clarify the basic molecular mechanisms of pituitary adenoma formation and to identify new tumor-related markers [33].

In a different study, Zhan et al. [29] have separated by 2D a total of 1000 proteins; 135 protein spots that represent 111 proteins were characterized with MS (96 spots for MALDI-TOF and 39 spots for LC-ESI-Q-IT). The characterized proteins include pituitary hormones, cellular signals, enzymes, cellular-defense proteins, cell-structure proteins, transport proteins, etc. These proteins were located in the cytoplasm, cellular membrane, mitochondria, endoplasmic reticulum, nucleus, ribonucleosome, and extracellular fractions, or were secreted in plasma. The identified proteins contribute to a functional profile of the pituitary adenoma proteome [29].

In 2010, Zhan et al. [34] found a series of pituitary adenoma proteomic expression data, which include 111 proteins identified from a human pituitary NF adenoma tissue, 56 DEPs; from human pituitary NF adenoma tissues and from prolactinoma tissues; 9 nitroproteins and 3 nitroprotein–protein complexes from a human pituitary NF adenoma tissue; and 8 nitroproteins from a pituitary control tissue. There is a pressing need to clarify the significant signaling pathway networks that involve those pituitary adenoma proteins, DEPs, and nitroproteins, in order to clarify and to better understand—at a molecular level—the pituitary adenoma pathogenesis. The knowledge of significant signaling pathway networks will provide important clues and clear directions for an in-depth investigation of pituitary adenomas, for the discovery of tumor biomarkers, and for the development of efficacious therapeutic agents [34].

The study of Zhan et al. [34] used bioinformatics pathway analysis—namely, IPA—to reveal for the first time the significant signaling pathways and networks that are associated with pituitary adenomas. The IPA system is an extensively used pathway analysis system that includes a large-scale knowledge base. IPA can identify statistically significant signaling pathway networks by analyzing the omic data in those numerous canonical pathway databases [34].

Three types of proteomic data were used—pituitary adenoma protein mapping, comparative proteomic, and nitroproteomic. Protein-mapping data were obtained with a 2DGE-arrayed pituitary adenoma proteome, followed by MS characterization of the proteins. Comparative-proteomic data were obtained from 2DGE-arrayed adenoma and control proteome images, followed by MS characterization of DEPs. Nitroproteomic data include those endogenous proteins that were nitrated at a tyrosine residue.

The comparative-proteomics data of Zhan et al. demonstrate that some components of mitochondrial complexes are significantly upregulated in pituitary adenomas relative to controls. Figure 9.1 shows the ERK/MAPK signaling pathway. The pathway analyses of our pituitary adenoma proteomic data clearly demonstrate that MAPK signaling pathways are involved in pituitary tumorigenesis.

Compared to human pituitary controls, in pituitary adenomas, the FYN (FYN oncogene related to SRC, FGR, and YES) was upregulated (4-fold), 14-3-3 protein downregulated (44-fold), HSPB1 (heat shock 27 kDa protein 1) downregulated (5-fold), and PPP2R2A (protein phosphatase 2 regulatory subunit B α-isoform) downregulated (8-fold), within the ERK/MAPK signaling pathway system [33].

Of the complicated pathway networks described above, several signaling pathways and networks were found to be significantly associated with pituitary adenoma, including mitochondria dysfunction, oxidative stress, cell-cycle deregulation, and the MAPK signaling system. These pathway network deregulations function in pituitary adenoma. The results demonstrate that mitochondria dysfunctions, oxidative stress, cell-cycle deregulation, and the MAPK signaling system are significantly associated with pituitary adenoma pathogenesis. Those data could provide biomarkers and could lead to the development of novel efficacious targets to treat pituitary adenomas. Therefore, changes of molecular phenotype are demonstrated to occur in pituitary tumorigenesis. Proteomics and transcriptomics were used to reveal pathological mechanisms and to discover biomarkers for molecular classification, diagnosis, and treatment.

Along with the aforementioned techniques, there are others that range from the ones that have been used for the past few decades. These range from *in situ* hybridization to many newly developed techniques, such as comparative-genomic hybridization, laser capture microdissection (LCM), and RNA interference technology. These techniques are being widely approached in Kontogeorgos et al.'s *Molecular Pathology of the Pituitary.*

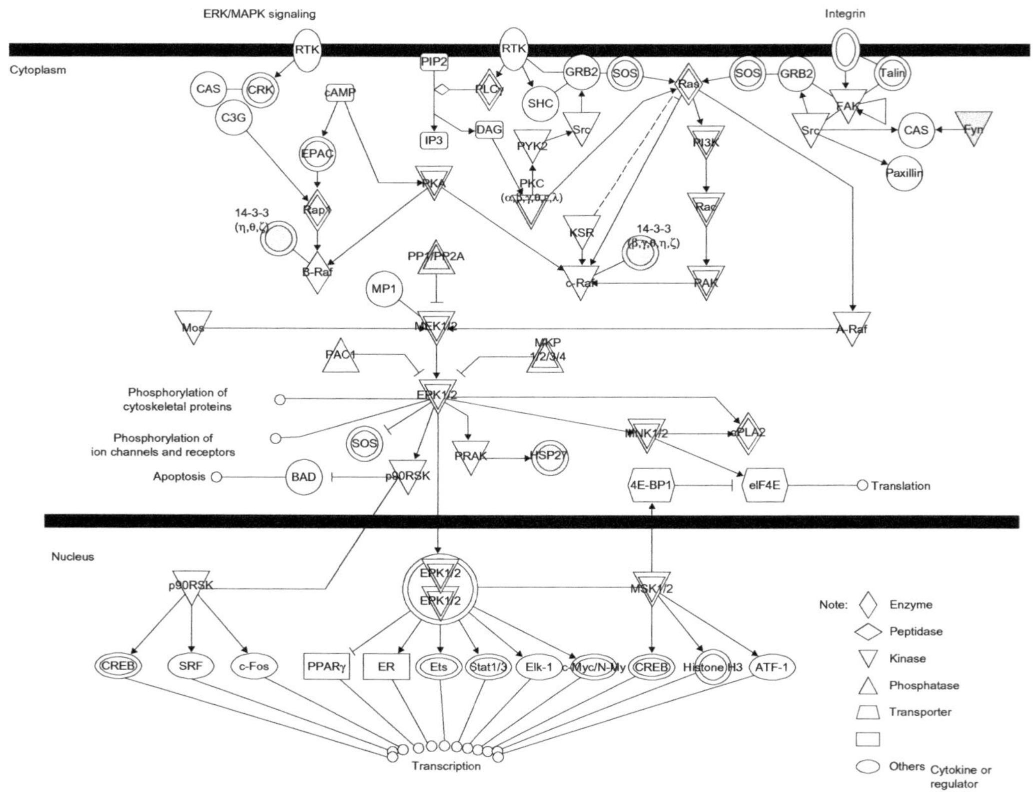

Figure 9.1 Comparative proteomics in pituitary adenomas ERK/MAPK signaling pathway that is involved with pituitary adenoma comparative-proteomic data. Red label = upregulated; green label = downregulated. The various shapes of nodes denote the different functions. A duplicated shape means this node contains multiple components. An arrow denotes the pathway direction. A line with a small circle denotes a biological result. (For interpretation of the references to color in this figure legend, the reader is referred to the web version of this book.)
Source: Republished with permission from [34].

In Situ Hybridization

Different forms of hybridization, such as *in situ* hybridization, have provided major insights into molecular mechanisms and disease development including studies in the pituitary. *In situ* hybridization represents a powerful technique used in molecular pathology.

A wide variety of mRNA transcripts have been identified in pituitary adenomas by *in situ* hybridization. These have ranged from transcripts for secretory granule protein messenger RNAs, such as the chromogranin/secretogranin family, to hypothalamic hormones and hormone receptors, as well as transcription factors and galectin-3 expression in normal and neoplastic pituitaries. Therefore, further insight into the pathogenesis of pituitary adenomas has been provided.

Laser Capture Microdissection

The acquisition of homogenous or pure cell populations for cell biologic and molecular analyses has been a difficult challenge for a long time, and many approaches have been tried; however, the development of LCM has provided a rapid and efficient method to capture pure cell populations for molecular and other studies. Kontogeorgos et al. [35] published a study in which they obtained pure populations of pituitary FSCs and used these for molecular studies that have provided new insights into the role of these cells in pituitary function.

RNA Interference

Very few studies have applied RNA interference technology to the pituitary. In one study [34], siRNA was used for galectin-3 to show that this protein was important for pituitary cell proliferation and apoptosis.

The development of high-throughput techniques has been applied to tissue analysis with the use of high-density arrays made up of 100 or more tissue samples, which provides an efficient method to validate and analyze molecular studies. However, nowadays fluid samples (serum, plasma) are also to be considered. The last acquisitions are on other tumors regarding the capacity of a large number of probes and analytes (cytokines, growth factors) by xMAP array—Luminex.

Our results [36] demonstrate that cytokine and angiogenic factor levels are closely linked to the pituitary tumoral behavior. The expression of cytokines/angiogenic factors is strongly related to tumor invasiveness, and in this way, it may act as a supporting factor for pituitary tumor expansion. Luminex xMAP might be a suitable tool for the evaluation of tumoral development. Increased levels of IL-1β, IL-6, and TNFα were found in 27%, 32%, and 41%, respectively. Cytokine expression was significantly higher in invasive pituitary adenomas (86%) compared to noninvasive ones (7%). TNFα level was up to 1.8-fold higher than the control, while IL-6 and

IL-1β were 4- and 4.7-fold higher. We have noticed a positive correlation between the cytokine level and tumoral invasiveness [35].

Growth factor values obtained by xMAP arrays were comparable with the outline obtained by ELISA tests. The mean value for soluble VEGF in patients versus the control was 1.6-fold in Luminex and 2-fold in ELISA analysis. Soluble bFGF mean value was 1.2-fold higher when Luminex technology was used, and 1.6-fold in ELISA quantification.

Taking into account the tendency toward individualized disease treatment for patients, there is an increasing need to understand the functioning mechanisms of pituitary adenomas, from general to particular, from a whole to a part, from organ to tissue, and from cellular to molecular.

References

1. The Innovative Medicines Initiative (IMI) (2006) Strategic research agenda "Creating Biomedical R&D Leadership for Europe to Benefit Patients and Society," www.ebe-biopharma.org/documents/IMI/imi_strategicresagenda.pdf.
2. Atkinson, A.J., Colburn, W.A., Degruttola, V.G., Demets, D.L., Downing, G.J., Hoth, D.F., et al. (2001) Biomarkers and surrogate endpoints: Preferred definitions and conceptual framework. *Clin. Pharmacol. Ther.*, 69(3): 89–95.
3. Zhan, X. and Desiderio, D.M. (2010) Signaling pathway networks mined from human pituitary adenoma proteomics data. *BMC Med. Genomics*, 3: 13.
4. Tainsky, M.A. (2009) Genomic and proteomic biomarkers for cancer: A multitude of opportunities. *Biochim. Biophys. Acta*, 1796(2): 176–93.
5. Ley, T.J., Mardis, E.R., Ding, L., Fulton, B., Mclellan, M.D., Chen, K., et al. (2008) DNA sequencing of a cytogenetically normal acute myeloid leukaemia genome. *Nature*, 456(7218): 66–72.
6. Ding, L., Getz, G., Wheeler, D.A., Mardis, E.R., McLellan, M.D., Cibulskis, K., et al. (2008) Somatic mutations affect key pathways in lung adenocarcinoma. *Nature*, 455: 1069–75.
7. Leary, R.J., Lin, J.C., Cummins, J., Boca, S., Wood, L.D., Parsons, D.W., et al. (2008) Integrated analysis of homozygous deletions, focal amplifications, and sequence alterations in breast and colorectal cancers. *Proc. Natl. Acad. Sci. U.S.A.*, 105: 16224–16229.
8. Okuda, S., Yamada, T., Hamajima, M.A., Itoh, M., Katayama, T., Bork, P., et al. (2008) KEGG Atlas mapping for global analysis of metabolic pathways. *Nucleic Acids Res.*, 36: W423–26.
9. Segota, I., Bartonicek, N. and Vlahovicek, K. (2008) MADNet: Microarray database network web server. *Nucleic Acids Res.*, 36: W332–35.
10. Hoque, M.O., Feng, Q., Toure, P., Dem, A., Critchlow, C.W., Hawes, S.E., et al. (2006) Detection of aberrant methylation of four genes in plasma DNA for the detection of breast cancer. *J. Clin. Oncol.*, 24: 4262–69.
11. Muraki, Y., Banno, K., Yanokura, M., Kobayashi, Y., Kawaguchi, M., Nomura, H., et al. (2009) Epigenetic DNA hypermethylation: Clinical applications in endometrial cancer. *Oncol. Rep.*, 22(5): 967–72.
12. Lawrie, C.H., Gal, S., Dunlop, H.M., Pushkaran, B., Liggins, A.P., Pulford, K., et al. (2008) Detection of elevated levels of tumor-associated microRNAs in serum of patients with diffuse large B-cell lymphoma. *Br. J. Haematol.*, 141: 672–75.

13. Roessler, M., Rollinger, W., Mantovani-Endl, L., Hagmann, M.L., Palme, S., Berndt, P., et al. (2006) Identification of PSME3 as a novel serum tumor marker for colorectal cancer by combining two-dimensional polyacrylamide gel electrophoresis with a strictly mass spectrometry-based approach for data analysis. *Mol. Cell. Proteomics*, 5: 2092–101.
14. Martosella, J., Zolotarjova, N., Liu, H., Nicol, G. and Boyes, B.E. (2005) Reversed-phase high performance liquid chromatographic prefractionation of immunodepleted human serum proteins to enhance mass spectrometry identification of lower-abundant proteins. *J. Proteome Res.*, 4: 1522–37.
15. Faca, V., Coram, M., Phanstiel, D., Glukhova, V., Zhang, Q., Fitzgibbon, M., et al. (2006) Quantitative analysis of acrylamide labeled serum proteins by LC-MS/MS. *J. Proteome Res.*, 5: 2009–18.
16. Andreev, V.P., Li, L., Rejtar, T., Li, Q., Ferry, J.G. and Karger, B.L. (2006) New algorithm for 15N/14N quantitation with LC-ESI-MS using an LTQ-FT mass spectrometer. *J. Proteome Res.*, 5: 2039–45.
17. Merchant, M. and Weinberger, S.R. (2000) Recent advancements in surface-enhanced laser desorption/ionization-time of flight-mass spectrometry. *Electrophoresis*, 21: 1164–77.
18. Petricoin, E.F., Zoon, K.C., Kohn, E.C., Barrett, J.C. and Liotta, L.A. (2002) Clinical proteomics: Translating benchside promise into bedside reality. *Nat. Rev. Drug Discov.*, 1: 683–95.
19. Drake, R.R., Schwegler, E.E., Malik, G., Diaz, J., Block, T., Mehta, A., et al. (2006) Lectin capture strategies combined with mass spectrometry for the discovery of serum glycoprotein biomarkers. *Mol. Cell. Proteomics*, 5: 1957–67.
20. Hanas, J.S., Hocker, J.R., Cheung, J.Y., Larabee, J.L., Lerner, M.R., Lightfoot, S.A., et al. (2008) Biomarker identification in human pancreatic cancer sera. *Pancreas*, 36(1): 61–9.
21. Haab, B.B. (2005) Antibody arrays in cancer research. *Mol. Cell. Proteomics*, 4: 377–83.
22. Sanchez-Carbayo, M., Socci, N.D., Lozano, J.J., Haab, B.B. and Cordon-Cardo, C. (2006) Profiling bladder cancer using targeted antibody arrays. *Am. J. Pathol.*, 168: 93–103.
23. Qian, X., Scheithauer, B.W., Kovacs, K. and Lloyd, R.V. (2005) DNA microarrays: Recent developments and applications to the study of pituitary tissues. *Endocrine*, 28(1): 49–56.
24. Farrell, W.E. (2006) Pituitary tumours: Findings from whole genome analyses. *Endocr. Relat. Cancer*, 13(3): 707–16.
25. Jiang, Z., Gui, S. and Zhang, Y. (2010) Analysis of differential gene expression by fiber-optic BeadArray and pathway in prolactinomas. *Endocrine*, 38(3): 360–68.
26. Evans, C.O., Moreno, C.S., Zhan, X., McCabe, M.T., Vertino, P.M., Desiderio, D.M., et al. (2008) Molecular pathogenesis of human prolactinomas identified by gene expression profiling, RT-qPCR, and proteomic analyses. *Pituitary*, 11(3): 231–45.
27. Moreno, C.S., Evans, C.O., Zhan, X., Okor, M., Desiderio, D.M. and Oyesiku, N.M. (2005) Novel molecular signaling and classification of human clinically nonfunctional pituitary adenomas identified by gene expression profiling and proteomic analyses. *Cancer Res.*, 65(22): 10214–10222.
28. Galland, F., Lacroix, L., Saulnier, P., Dessen, P., Meduri, G., Bernier, M., et al. (2010) Differential gene expression profiles of invasive and non-invasive non-functioning pituitary adenomas based on microarray analysis. *Endocr. Relat. Cancer*, 17(2): 361–71.
29. Zhan, X. and Desiderio, D.M. (2003) A reference map of a human pituitary adenoma proteome. *Proteomics*, 3(5): 699–713.
30. Zhan, X. and Desiderio, D.M. (2003) Heterogeneity analysis of the human pituitary proteome. *Clin. Chem.*, 49(10): 1740–51.

31. Beranova-Giorgianni, S., Giorgianni, F. and Desiderio, D.M. (2002) Analysis of the proteome in the human pituitary. *Proteomics*, 2(5): 534–42.
32. Desiderio, D.M. and Zhan, X. (2003) The human pituitary proteome: The characterization of differentially expressed proteins in an adenoma compared to a control. *Cell. Mol. Biol.*, 49(5): 689–712.
33. Zhan, X., Evans, C.O., Oyesiku, N.M. and Desiderio, D.M. (2003) Proteomics and transcriptomics analyses of secretagogin down-regulation in human non-functional pituitary adenomas. *Pituitary*, 6(4): 189–202.
34. Zhan, X. and Desiderio, D.M. (2010) Signaling pathway networks mined from human pituitary adenoma proteomics data. *BMC Med. Genomics*, 3: 13.
35. Kontogeorgos, G. and Kovacs, K. (2004) Molecular Pathology of the Pituitary. Karger, Basel.
36. Tanase, C., Raducan, E., Mihai, S., Georgescu, M., Ionita, A., Poroschianu, M., et al. (2009) Serum cytokines and growth factors levels in pituitary adenomas behaviour using Xmap technology. *Virchows Arch.*, 455: S383.

10 Therapeutic Approaches: Drug, Surgical, and Radiotherapy

Nowadays, the treatment of pituitary adenomas is handled in a multimodal manner. Despite being considered as benign, excess hormone secretion or hypopituitarism implies a major impact upon health, with increased cardiovascular events and mortality. With the exception of pituitary incidentalomas, with increasing incidence, pituitary adenomas (secreting or non-functional) requires specific intervention. The detailed characterization of adenoma is a prerequisite of successful treatment. Therefore, the clinical picture, complications (optic pathway compression, pituitary apoplexy), detailed imaging and the whole set of endocrine evaluation of the adenoma impact upon treatment decision [1].

Clinically, aggressive pituitary adenomas are difficult to manage. A significant number of patients with these adenomas have invasion of bone, dura, and/or adjacent structures, thus complicating or precluding their resection. Patients with aggressive pituitary adenomas present with problems caused largely by incomplete resection, recurrence, and associated morbidity.

Surgical Treatment

The main principles regarding the surgical indication in pituitary adenomas (except for prolactinomas) recommend surgical resection as the main treatment in tumors that are endocrinologically active, in endocrinologically inactive tumors accompanied by mass effect through the compression and/or the infiltration of the adjacent tissues, as well as in tumors accompanied by pituitary insufficiency or in those whose progression is monitored [2]. In the last decade, pituitary surgery developed by being less invasive (via transsphenoidal approach), more complete, owing to high resolution optics coupled with neuronavigation systems and intraoperative imaging procedures. Completeness of resection and preservation of normal pituitary as well as decompression of optic pathway are the main neurosurgical targets, as well as performing recursion of the hormonal values at their physiological levels, the obtaining of the biochemical cure or remission, in the case of secreting tumors, and recursion at the normal pituitary parameters, in the case of pituitary insufficiency. Not less important, surgery provide pathologists and scientists with tumor tissue for a better characterization of their biology. Therefore, the postsurgical monitoring is realized both by imagistic control and by endocrine monitoring. Due to wide variation in these data, the first line treatment is still under debate [2,3,4].

Drug Treatment

In prolactinomas and GH-PRL secreting adenomas, dopamine agonists are the first line therapy, with improvement/normalization of PRL levels and restore of pituitary

Molecular Pathology of Pituitary Adenomas. DOI: 10.1016/B978-0-12-415830-6.00010-X function, as well as tumor shrinkage. Real cure rate in prolactinomas is limited, stop of dopamine agonists being associated with relapse of secretion and tumor reexpansion. In balance, surgery in microprolactinomas is highly successful [5].

For the GH secreting tumors in acromegaly, the first line treatment is still under debate. More and more patients are treated first by somatostatin analogues (SMSa), while surgery goes afterwards.

Absence of normalization of the hormonal tumor markers (PRL or GH and IGF1) as well as lack of tumor shrinkage under medical treatment defines tumor resistance. In order to normalize IGF1, the peripheral marker of active acromegaly, an analogue of GH with receptor blocking properties- pegvisomant – has been developed. New data support the hypothesis that addition of Pegvisomant to somatostatin analogues improves global outcome in treating acromegaly, if surgery fails to cure. [1]. Other tumors, as nonfunctioning adenomas, Cushing or thyrotropinomas are again submitted to surgery as first line, while dopamine or SMSa resistant adenomas are submitted to surgery after documented failure to control secretion or to shrink the tumor.

Treatment with Temozolomide

Beyond the three categories of treatment, in severe cases which are close to malignancy in terms of tumor biology, other additional resources can be used. Temozolomide, drugs targeting angiogenesis or gene therapy are in the front line research and treatment in trials involving patients with pituitary adenomas. Temozolomide, an orally administered alkylating agent, is used to treat malignant gliomas. Recent reports also have documented its efficacy in the treatment of pituitary adenomas and carcinomas. Temozolomide methylates DNA and thereby exhibits an antitumor effect. O6-methylguanine-DNA methyltransferase (MGMT), a DNA repair enzyme, removes alkylating adducts induced by temozolomide, counteracting its effects [6].

In 2006, the first treatment with temozolomide for pituitary adenoma was realized. Since then, several researchers have reported treatments with temozolomide. Syro et al. [7] and Kovacs et al. [8,9], published the only morphologic comparisons of pathology before and after temozolomide treatment.

The temozolomide-treated tumor had better differentiation and consisted of larger cells, exhibited fewer mitoses, and had a lower Ki-67 LI. Tumor hemorrhage, necrosis, focal fibrosis, and neuronal transformation also were observed [10], and MGMT immunoexpression was lacking entirely.

Kovacs et al. [8] were the first to report the inverse relation between MGMT immunoexpression and temozolomide response, which they found in two patients with aggressive adenomas. The reports of Kovacs et al. [8] and McCormack et al. [11] underscored the inverse relation between MGMT immunoexpression and the efficacy of temozolomide therapy.

In recurrent, aggressive pituitary adenomas and carcinomas that are resistant to multimodality therapy, radiotherapy, pharmacologic manipulation, and multiple conventional chemotherapy, the assessment of MGMT immunoexpression may serve as a predictor of the response to temozolomide. If MGMT expression is low, then a dramatic response can be anticipated.

Temozolomide has demonstrated value in the treatment of aggressive pituitary adenomas and carcinomas. The clinical and radiologic responses are encouraging. Clearly, there is an inverse correlation between MGMT immunoexpression and the therapeutic response to temozolomide. Targeted modulation of MGMT may be useful in patients who otherwise may not respond to temozolomide therapy [7].

Therapeutic Approaches—Targeting Signaling Molecules

Many studies focus upon the current preclinical knowledge regarding the therapeutic implication of signaling molecules that are involved in cell proliferation, angiogenesis, the cell cycle, and so on. Modulators of the PI3K–AKT pathway have been focusing mainly on inhibitors and other compounds targeting PI3K, PDK1, AKT, and HSP90. The use of modulators for the PI3K–AKT pathway as potential targeted anticancer agents is still at an early stage but has already entered clinical trials. Blocking the PI3K–AKT pathway at different nodes might yield different antitumor activities and therapeutic windows [12,13]. The complex network in which the PTTG is involved, regulating signaling pathways for growth factors and GHs, allows pharmacological intervention using receptor blockade or key intracellular molecule inhibition. Pituitary adenoma therapy could benefit from the inhibition of EGFR-mediated PTTG1 expression in FSCs by intracellular blockade of paracrine signaling [14].

Gene therapy techniques that specifically target endogenous PTTG1 expression in tumors may serve as a useful investigative therapeutic tool [15]. The observed cell cycle deregulation in pituitary disease is another important therapeutic target. The frequent overexpression of cyclins and inactivation of cell cycle inhibitors such as INK4 proteins suggest that CDK hyperactivation is a common theme in pituitary neoplasia. Several small CDK inhibitors are now being evaluated for cancer therapy in many different tumor types [16,17]. Although these drugs have not been clinically tested in pituitary tumors, preclinical studies suggest that CDK inhibitors may be effective for treating pituitary diseases, at least in individuals with cell cycle mutations that specifically affect this pathway [18].

As mentioned above, the main medical treatment currently available for prolactinomas and acromegaly are the dopamine and somatostatin analogues, respectively. They act both on tumor secretion, decreasing the PRL/GH levels, as well as on tumor size, leading to variable tumor shrinkage [19]. Their failure reflects tumor resistance and requires further non-specific approaches (surgery and radiotherapy). Therefore, morphological assessment of the SSTR profile can predict the responsiveness and validate the effectiveness of treatment with SST analogs [20,21]. More potent and broader-spectrum SST analogs are likely to play an increasing role in the treatment of tumors, in which the MAPK pathway is overactivated [22].

Many clinically non-functioning adenomas are immunopositive for glycoproteic hormones but targeting the GnRH receptor was not beneficial for the tumor control [23]. Because pituitary transformation resides in several processes based on key signaling deregulation, we do not rule out that a therapeutic approach first has to personalize the molecular pattern of the pituitary adenoma and then develop a multitargeting therapy.

Radiotherapy Treatment

While the first and second place are disputed, almost everyone agrees upon radiotherapy as last treatment resource. Severe hypopituitarism, delayed response and increased rate of cerebrovascular events are only few of the drawbacks of the method. In selected cases with aggressive biology, it is important to use this ultimate resource to limit the tumor expansion/relapse. Various treatment options are available, as stereotaxic radiotherapy or gamma knife surgery improved in the last decade [3,4].

Future Perspectives and Therapeutic Implications in Pituitary Adenomas

We can foresee that SST analogs and chimerical constructs will enter the routine clinical medication of pituitary tumors [24]. Future clinical studies will select specific inhibitors for the PI3K–AKT pathway that could simultaneously inhibit the proliferation and growth of tumor cells and provide the best therapeutic results. For personalized therapy, studies regarding specific genetic and epigenetic alterations associated with pituitary adenomas will direct therapeutic approaches toward the optimum outcome.

As the involvement of pituitary oncogene PTTG1 is not a matter of debate, we can predict that in the near future, new classes of genetic therapeutic drugs will target this oncogene. We cannot rule out the possibility that cell cycle circuits involved in the development of pituitary syndromes and tumors will provide important molecules associated to pituitary tumor development in the future.

References

1. Buchfelder, M. (2005) Treatment of pituitary tumors. *Endocrine*, 28(1): 67–75.
2. Fahlbusch, R., Keller, B., Ganslandt, O., Kreutzer, J. and Nimsky, C. (2005) Transsphenoidal surgery in acromegaly investigated by intraoperative high-field magnetic resonance imaging. *Eur. J. Endocrinol.*, 153(2): 239–48.
3. Ghostine, S., Ghostine, M.S. and Johnson, W.D. (2008) Radiation therapy in the treatment of pituitary tumors. *Neurosurg. Focus*, 25(5): 1–5.
4. Molitch, M.E. and Grossmann, A.B. (2008) Pituitary radiotherapy. *Pituitary*, 12: 1–2.
5. Ivan, G., Syiegeti-Csucy, N. and Olah, M. (2005) Treatment of pituitary tumors: Dopamine agonists. *Endocrine*, 28(1): 101–10.
6. Syro, L.V., Kovacs, K. and Scheithauer, B.W. (2011) Reply to treatment of pituitary neoplasms with temozolomide: A review. *Cancer*, 117: 4102.

7. Syro, L.V., Uribe, H., Penagos, L.C., Ortiz, L.D., Fadul, C.E., Horvath, E., et al. (2006) Antitumour effects of temozolomide in a man with a large, invasive prolactin-producing pituitary neoplasm. *Clin. Endocrinol. (Oxf.)*, 65: 552–53.
8. Kovacs, K., Horvath, E., Syro, L.V., Uribe, H., Penagos, L.C., Ortiz, L.D., et al. (2007) Temozolomide therapy in a man with an aggressive prolactin-secreting pituitary neoplasm: Morphological findings. *Hum. Pathol.*, 38: 185–89.
9. Kovacs, K., Scheithauer, B.W., Lombardero, M., McLendon, R.E., Syro, L.V., Uribe, H., et al. (2008) MGMT immunoexpression predicts responsiveness of pituitary tumors to temozolomide therapy. *Acta Neuropathol.*, 115: 261–62.
10. Agarwala, S.S. and Kirkwood, J.M. (2000) Temozolomide, a novel alkylating agent with activity in the central nervous system, may improve the treatment of advanced metastatic melanoma. *Oncologist*, 5: 144–51.
11. McCormack, A.I., McDonald, K.L., Gill, A.J., Clark, S.J., Burt, M.G., Campbell, K.A., et al. (2009) Low O6-methylguanine-DNA methyltransferase (MGMT) expression and response to temozolomide in aggressive pituitary tumours. *Clin. Endocrinol. (Oxf.)*, 71: 226–33.
12. Garcia-Echeverria, C. and Sellers, W.R. (2008) Drug discovery approaches targeting the PI3K/AKT pathway in cancer. *Oncogene*, 27(41): 5511–26.
13. Carnero, A., Blanco-Aparicio, C., Renner, O., Link, W. and Leal, J.F. (2008) The PTEN/PI3K/AKT signalling pathway in cancer, therapeutic implications. *Curr. Cancer Drug Targets*, 8(3): 187–98.
14. Vlotides, G., Cruz-Soto, M., Rubinek, T., Eigler, T., Auernhammer, C.J. and Melmed, S. (2006) Mechanisms for growth factor-induced pituitary tumor transforming gene-1 expression in pituitary folliculostellate TtT/GF cells. *Mol. Endocrinol.*, 20(12): 3321–35.
15. Vlotides, G., Eigler, T. and Melmed, S. (2007) Pituitary tumor-transforming gene: Physiology and implications for tumorigenesis. *Endocr. Rev.*, 28(2): 165–86.
16. Malumbres, M., Pevarello, P., Barbacid, M. and Bischoff, J.R. (2008) CDK inhibitors in cancer therapy: What is next? *Trends Pharmacol. Sci.*, 29(1): 16–21.
17. Perez de Castro, I., de Carcer, G. and Malumbres, M. (2007) A census of mitotic cancer genes: New insights into tumor cell biology and cancer therapy. *Carcinogenesis*, 28(5): 899–912.
18. Sotillo, R., Renner, O., Dubus, P., Ruiz-Cabello, J., Martín-Caballero, J., Barbacid, M., et al. (2005) Cooperation between Cdk4 and p27kip1 in tumor development: A preclinical model to evaluate cell cycle inhibitors with therapeutic activity. *Cancer Res.*, 65(9): 3846–52.
19. Heaney, A.P. and Melmed, S. (2004) Molecular targets in pituitary tumours. *Nat. Rev. Cancer*, 4(4): 285–95.
20. Kontogeorgos, G. (2005) Classification and pathology of pituitary tumours. *Endocrine*, 28(1): 27–35.
21. Kontogeorgos, G. (2006) Predictive markers of pituitary adenoma behavior. *Neuroendocrinology*, 83(3–4): 179–88.
22. Hubina, E., Nanzer, A.M., Hanson, M.R., Ciccarelli, E., Losa, M., Gaia, D., et al. (2006) Somatostatin analogues stimulate p27 expression and inhibit the MAP kinase pathway in pituitary tumours. *Eur. J. Endocrinol.*, 155(2): 371–79.
23. Naor, Z. (2009) Signalling by G-protein-coupled receptor (GPCR): Studies on the GnRH receptor. *Front. Neuroendocrinol.*, 30(1): 10–29.
24. Tanase, C.P., Neagu, M. and Albulescu, R. (2009) Key signaling molecules in pituitary tumors. *Expert Rev. Mol. Diagn.*, 9(8): 859–77.

www.ingramcontent.com/pod-product-compliance
Ingram Content Group UK Ltd.
Pitfield, Milton Keynes, MK11 3LW, UK
UKHW020124290926
15471UKWH00003B/265

9 780323 165358